Better Homes and Gardens®

STEP-BY-STEP
BASIC
WIRING

BETTER HOMES AND GARDENS® BOOKS
Editor: Gerald M. Knox
Art Director: Ernest Shelton

Building and Remodeling Editor: Noel Seney
Building Books Editor: Larry Clayton

Associate Art Directors: Neoma Alt West,
    Randall Yontz
Copy and Production Editors: David Kirchner,
    Lamont Olson, David Walsh
Assistant Art Director: Harijs Priekulis
Senior Graphic Designer: Faith Berven
Graphic Designers: Linda Ford,
    Sheryl Veenschoten, Tom Wegner

Editor in Chief: James A. Autry
Editorial Director: Neil Kuehnl
Group Administrative Editor: Duane Gregg
Executive Art Director: William J. Yates

Step-By-Step Basic Wiring
Editors: Noel Seney, Larry Clayton
Copy and Production Editor: Lamont Olson
Graphic Designer: Thomas Wegner
Contributing Writer: James A. Hufnagel
Technical Consultants: James Downing;
    Ron Tesdell, Tesdell Electric
Drawings: Carson Ode

Acknowledgments
Our appreciation goes to the following
companies and associations for contrib-
uting material helpful in the preparation
of this book:
Broan Manufacturing Company, Inc. (pages 51-53)
Graybar Electric Company, Inc.
Honeywell (pages 34, 45)
Leviton Manufacturing Company, Inc.
National Fire Protection Association
NuTone Division, Scovill Mfg. Co. (pages 54-55)
Raco, Inc.

CONTENTS

Introduction   4

Getting to Know Your System   6

Anatomy of a circuit... Fuses or breakers protect against fire...
Amps x Volts = Watts... Grounding protects against shock

Tools for Electrical Work   10

Solving Electrical Problems   12

Anatomy of a breaker box, 12... Troubleshooting tripped circuit breakers,
13... Anatomy of a fuse box, 14... Troubleshooting blown fuses, 14...
Replacing plugs, cords, and lamp sockets, 16... Replacing switches and
receptacles, 20... Troubleshooting incandescent fixtures, 22 ... Testing
switches, 23... Troubleshooting fluorescent fixtures, 24... Repairing a silent
door chime, 26

Making Electrical Improvements   28

Anatomy of a lamp, 28... Wiring your own lamps, 29... Hanging new
fixtures and track lights, 31... Installing dimmer switches, 32 ... Installing a
timed-setback thermostat, 34... Wiring receptacles, fixture outlets, and
switches to existing circuits, 35... Wiring a smoke detector, 45 ... Installing
a ground fault circuit interrupter, 46... Running wires underground, 48...
Installing a bathroom ventilating fan, 51... Installing a power attic fan,
52... Installing an intercom, 54... Adding circuits to your system, 56

Electrical Basics and Procedures   60

Selecting cable, 61... Electrical boxes, accessories, and how to install
them, 62... How to work with nonmetallic sheathed cable, 68 ... How to
work with conduit, 78... How to work with flexible armored cable (BX) and
flexible metallic conduit (Greenfield), 80, 82... Your switch and receptacle
options, 84... Making electrical connections, 86

Products You May Want to Try   90

Glossary   93

Index   95

INTRODUCTION

How do you feel about working with electricity? Many otherwise intrepid do-it-yourselfers would rather cozy up to a deadly snake. "Sure," they say, "the power's supposed to be off, but what if it's not? And couldn't a simple mistake set the house on fire?"

Electricity's mighty energy most certainly deserves respect, but then so does gasoline. Pour a few gallons into a car with bad brakes, get behind the wheel, and you're headed for trouble.

Neglecting your home's wiring can be every bit as devastating. Faulty electrical equipment claims thousands of lives and homes every year, in too many cases because someone didn't bother to make a simple repair.

To help clear away any phobias or misunderstandings you might have about electrical principles, this book begins with a series of visual analogies that represent how household wiring works and how two vital safety systems protect against fire and shock. These drawings also

introduce you to the primary electrical colors—black, white, and green—that enable you to decipher exactly what's going on at any point in your home's network of wires.

Next, we turn to those all-important repair jobs. Most of us feel some trepidation every time we remove the wall plate from a receptacle or drop a light fixture from the ceiling, but you needn't worry about shock as long as you observe these two cardinal rules:
- Always shut off power to the circuit you'll be working on, or the entire house if you're not sure which fuse or breaker controls the circuit.
- Double-check with a testing device to be absolutely sure the circuit is dead.

To help you remember, we've red-lettered power-off jobs and other important safety

information. When you see red, proceed with caution.

After you begin to understand how your home's electrical system works and have successfully completed a few repair jobs, you'll probably begin to think of some improvements you'd like to make. We present a dozen of them, everything from wiring your own custom lamps to hooking in a new circuit.

Try a few improvements and you'll discover just what a joy electrical work can be. Compared to carpentry, lawn-keeping, painting, and other around-the-house chores, wiring projects involve a minimum of time, trouble, and mess. In fact, most of the work takes place in your head—while you figure out what's to be done—and with your feet—as you travel to the service panel to cut and restore the power.

One big reason electrical jobs go so quickly lies with the highly modular components you'll learn about in our final major section, Electrical Basics and Procedures. It tells how to select the right equipment and how to interconnect all the elements correctly.

When you shop for wiring materials, look for the same UL symbol you've seen on appliances and other items. This means the equipment has been examined for safety defects by Underwriters' Laboratories, an independent testing organization. Using products that are not UL-listed may violate your local electrical code and possibly your home's fire insurance contract.

Incidentally, insurance policies typically say nothing one way or the other about fires caused by owner-installed wiring, but if you're thinking about taking on a major project, it's worth checking your policy's fine print.

Of course the best insurance against insurance hassles and the tragedies that cause them is to do the work right to begin with, and this, in electrical parlance, means working strictly "to code."

Working to Code

As you may already know, local and, in some cases, state laws, known as "codes," mandate what wiring materials you can use in your home and how they're to be installed. Ignoring these statutes not only risks a fine and an order to do the work over again, it also jeopardizes your family's safety.

Fortunately, modern codes are almost as standardized as the components you'll be working with. In the U.S., credit for this goes to the National Fire Protection Association, a non-profit, industry-wide agency that publishes the National Electrical Code (NEC).

The NEC is a voluminous "bible" of rules and regulations that covers just about any conceivable electrical situation. Although not law itself, the NEC serves as a model on which virtually all local codes are based. Some communities have simply adopted the national code as their own. Others are slightly more restrictive, but none permit practices that would be in violation of the national code.

The procedures shown in this book represent the editors' understanding of the 1978 NEC. (It's updated every three years.) If you're contemplating a major wiring job, check into both the NEC and local ordinances before you begin.

Technical bookstores often carry copies of the *National Electrical Code,* or write to the National Fire Protection Association, 470 Atlantic Avenue, Boston, MA 02210. Canadian residents can obtain a copy of the Canadian Electrical Code by writing to the Canadian Standards Association, 178 Rexdale Blvd., Rexdale, Ontario, Canada, M9W 1R2.

Realize, too, that if you're planning to extend an existing circuit or add a new one, you may need to apply to your community's building department for a permit, and arrange to have the work inspected before it's put into service. Local officials can tell you about this and whether you'll need a licensed electrician to "sign off" the job.

GETTING TO KNOW YOUR SYSTEM

Electrical energy, as you undoubtedly know, flows into and through your home via a network of wires. You've probably never seen most of these wires, though. They're wrapped with insulation, sometimes clad in metal, and usually concealed inside walls, floors, and ceilings.

Even if you could get a good look at a live, bare wire, you wouldn't see any action. That's because electricity consists of invisible particles of matter, called *electrons*, moving at the speed of light.

Electrons always flow in circles, known as *circuits*. To trace their route through one of your home's circuits, follow the trucks along the special sort of freeway illustrated here.

It all begins at your power supplier's generating plant, which pumps out charged electrons, illustrated by the black loads of the trucks. Overhead or underground wires bring power from the utility company's lines to your home's *service entrance,* then through a *meter* that keeps track of your household consumption.

Next, traffic proceeds to the *service panel,* which directs the flow to a series of circuits such as this simplified example. (Fuses or circuit breakers here stop traffic instantly if an accident occurs in any circuit. More about these on pages 8 and 12-15.)

Now the freeway narrows to two lanes, corresponding to the two wires needed for a single 120-volt circuit. One of them is typically covered in black insulation, identifying it as "hot" with charged traffic. The other wire always wears white insulation, signifying that it carries no charge.

Every potential exit from an electrical circuit is called an *outlet,* regardless of whether it consists of a *receptacle, switch,* or *light.* At our first exit, a receptacle outlet, electron traffic passes right on by because nothing is plugged in and turned on. But the moment you hook in an electrical device, charged electrons flow into it and do their work, as shown at the second receptacle. In the process, the electrons "dump" their charges. Now they begin a return trip back to the service panel.

Notice, too, what's happening down the road. Here charged electrons are being halted by a light switch in its off position. Flip on the switch, and the energy flow begins.

Back at the service panel, all empty, uncharged electrons return to the power company, or to the earth via a *ground.* More about this vital safety element on page 9.

ANATOMY OF A CIRCUIT

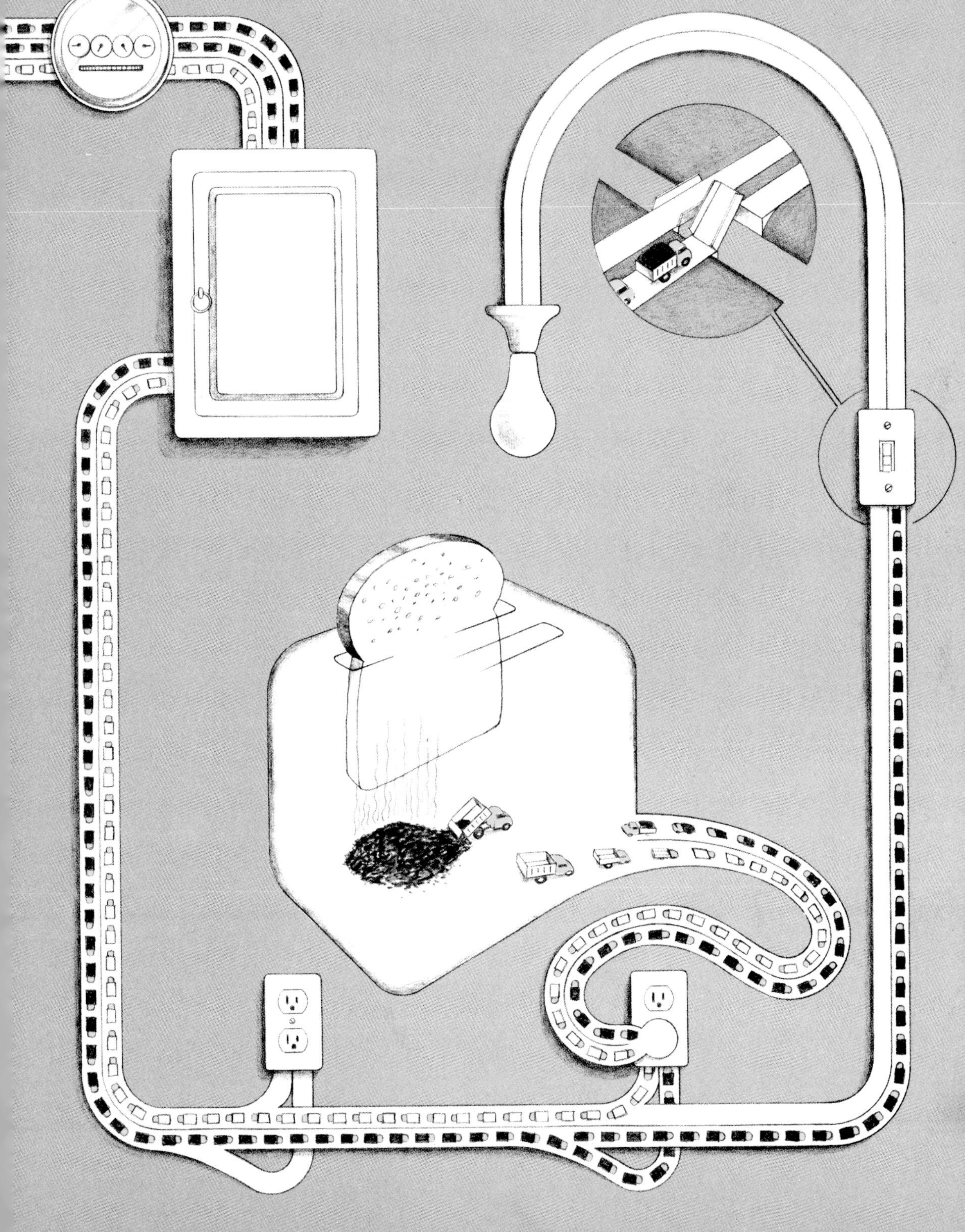

Fuses or Breakers Protect Against Fire

The freeway illustrated on the preceding page shows how traffic normally flows through an electrical circuit, but what if an accident occurs?

First, let's consider a head-on collision between our circuit's two "lanes." The result would be a *short circuit*, such as the one shown in sketch A. When a short occurs, traffic instantaneously piles up, and electricity's energy could touch off a fire.

That's why every properly wired household circuit includes a *fuse* or *circuit breaker* at the service panel. Either of these devices serves as a traffic cop. If a pileup occurs, it almost immediately shuts down the entire circuit.

Fuses and breakers also stop traffic if the circuit becomes *overloaded* with more current than it can handle. In this situation (see sketch B) there would be no head-on collision, but too much traffic would generate excess heat that might start a fire. For additional information about circuit breakers, fuses, shorts, and overloads, see pages 12-15.

Amps x Volts = Watts

The key to making good use of this very important electrical equation lies in your understanding of the terms used in it, so let's start there.

Amps represents the number of electrons flowing through a conductor. (This rate of flow can vary considerably, depending on the demand.) *Voltage* is the load of potential energy being carried by the electrons. And *wattage* measures the total energy consumed.

In our example (sketch C) we show two roadways on which trucks (electrons) are traveling. Both of these freeways have 120 volts of potential energy, so why is the low road twice as busy as the high road? Because one customer is using twice as much energy (wattage) as the other.

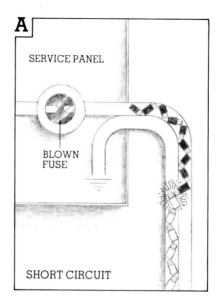

A

SERVICE PANEL

BLOWN FUSE

SHORT CIRCUIT

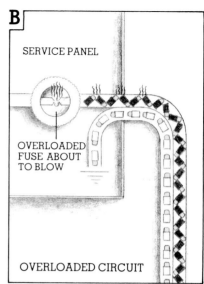

B

SERVICE PANEL

OVERLOADED FUSE ABOUT TO BLOW

OVERLOADED CIRCUIT

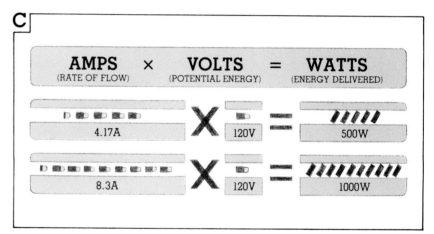

C

AMPS (RATE OF FLOW)	×	VOLTS (POTENTIAL ENERGY)	=	WATTS (ENERGY DELIVERED)
4.17A	X	120V	=	500W
8.3A	X	120V	=	1000W

One way to measure the amperage of our freeway circuit would be to count the number of trucks going by a certain point in one second. You'd have to be a fast counter because one amp stands for billions and billions of electrons. This is why the amps x volts = watts formula was devised.

Using it, you can compute the amount of energy required to operate any appliance or other item on a circuit. Smaller electrical devices, such as light bulbs, are commonly rated according to the wattage they draw. Bigger power users, such as appliances, often are given amp ratings. To convert from watts to amps, simply divide the wattage by the voltage.

To show how helpful the formula can be, let's look at an

example. Suppose you are thinking about buying an air conditioner rated at 9 amps and want to know if an existing circuit can handle it.

First check the circuit's fuse or breaker for its amp rating. In most cases this will be 15. Next, add up the wattages of every electrical "customer" on the circuit. Let's say the total is 600. Dividing this by the circuit's voltage tells you how many amps the circuit is already drawing. At 120 volts it would amount to 5 amps.

Now add this to the newcomer's demand — 5+9 — and you get 14, one amp less than the circuit's capacity.

Grounding Protects Against Shock

Head-on collisions and traffic jams aren't the only types of freeway incidents, nor are short circuits and overloads the only things that can go wrong with electrical traffic.

Examine drawing A, which shows an ungrounded circuit, and you'll see what can happen when some of our electron trucks get off at an other-than-normal exit, a situation that can occur inside a faulty motor, fixture, receptacle, or switch.

In this case a small amount of current is leaking, but because there's no collision, the fuse or breaker back at the service panel won't shut off the circuit. Some traffic, in effect, is stalled.

But electricity likes to keep moving. Pulled by the earth's magnetic force, it always seeks the shortest possible route to the ground. This means that if you are in contact with the earth and touch a device that's leaking current, all the stalled electrons will detour through your body.

Electrical shocks can amount to anything from a slight tingle to a fatal jolt, depending on how much current is leaking and how well you're grounded. If you happen to be touching a water pipe, or standing on a wet lawn, you could be seriously hurt or killed.

That is why today's codes, Underwriters' Laboratories, and common sense require that all equipment using substantial amounts of current be *grounded* with a third "lane" that safely detours stray current to the earth.

In drawing B we've represented the grounding path in green, the color commonly used in wiring diagrams. Some of the grounds in your home may well be covered in green or green and yellow insulation, although grounding wires needn't be insulated at all. Many are bare, and if the wires that run through your walls are metal-clad, the metal itself may be serving as the ground lane.

The important thing about any ground is that it follow an unbroken path to the earth. You're probably familiar with the three-blade plugs that most major appliances have. The third, rounded blade carries the ground and fits into a corresponding slot in newer receptacles. The receptacle itself and often the box that houses it are in turn grounded

back to the service panel. (Note: Only metallic boxes are grounded; nonmetallic boxes needn't be. Power tools with "double-insulated" plastic housings don't require grounding, either.)

(With older, two-slot receptacles, you may or may not get a safe ground with a special *adapter*. That depends on whether the receptacle and its box are grounded. To learn whether yours are, see page 21.)

Back at the service panel, the grounding wire for each circuit is connected to a *bus bar*, as are all the white neutral wires. A *system ground wire* then ties the bus bar to a metal water pipe, or a metal rod driven into the earth.

Why ground the white wires? This protects you and your home's electrical system against the enormous surge of current that might occur if the utility company's lines become short-circuited or are hit by lightning.

A newer code provision calls for another safety device in certain circuits. It's called a *ground fault circuit interrupter*. To learn what it does and how to install one, see pages 46 and 47.

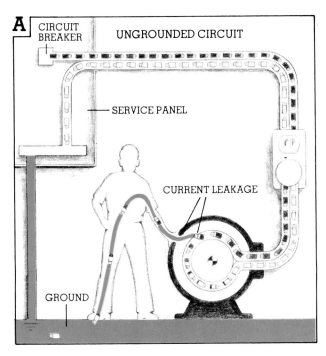

A CIRCUIT BREAKER UNGROUNDED CIRCUIT

SERVICE PANEL

CURRENT LEAKAGE

GROUND

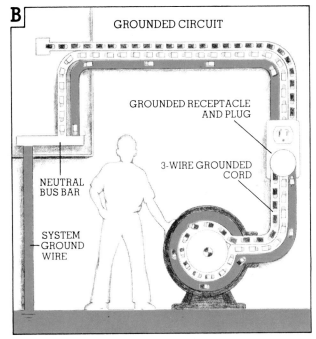

B GROUNDED CIRCUIT

GROUNDED RECEPTACLE AND PLUG

3-WIRE GROUNDED CORD

NEUTRAL BUS BAR

SYSTEM GROUND WIRE

TOOLS FOR ELECTRICAL WORK

You don't need an arsenal of specialized equipment to do the electrical projects in this book. In fact, for most repair jobs, you can get by with just a couple of screwdrivers, two or three pairs of pliers, and one or two of the testing devices shown here. For improvements, you need to add a few more items.

Long-nose and *linemen's pliers* are musts. The first help you curl wires into the loops needed for many electrical connections. Linemen's pliers handle heavier cutting and twisting jobs. *Side-cutting pliers* are handy when you have to snip wires in tight places.

A *utility knife* slices through insulation, drywall, and almost anything else that calls for a cutting edge. A *combination tool* does a variety of tasks—cuts and strips wires, sizes and cuts off screws, crimps connectors, and more. If your wiring is protected by cartridge-type fuses, a plastic *fuse puller* lets

you remove them without danger of electrical shock.

With testers, you have some options. The *continuity* type has a small battery and bulb so you can test switches, sockets, and fuses, but only with the power off. A *neon tester*, on the other hand, lights up only when current is flowing. It tells you whether an outlet is live and hooked up properly.

You'll need one of each of these, or you can invest in an inexpensive *voltmeter* instead. This one works with the power on or off, and also measures how much voltage you have at an outlet. If you'll be installing a number of new receptacles, you might also want a *receptacle analyzer*. Plug it in and indicator lights warn you of faulty connections. You can learn the same things with a neon tester or voltmeter, but it takes more time.

Thinking about running some new wiring? Regardless of how

you plan to do it, you'll probably need an *electric drill*, a *spade bit*, and maybe a *bit extension* to lengthen the bit's reach.

If codes permit nonmetallic sheathed cable in your area, use a *cable ripper* to cut open the plastic sheath quickly. This one also strips insulation from conductors. For conduit, you'll need a *tubing cutter* or a hacksaw, a *bender*, and a *fish tape* to pull wires through the run. With a pair of tapes, you can also sneak cables through finished walls and ceilings.

Some codes require soldered connections in house wiring, and low-voltage installations also are usually soldered. A *soldering gun* and a spool of rosin-core *solder* do this job.

Most of the gear shown here packs easily into a *tool pouch*. Electrical work keeps you on the move, and you'll appreciate the convenience of having anything you might need at your hip.

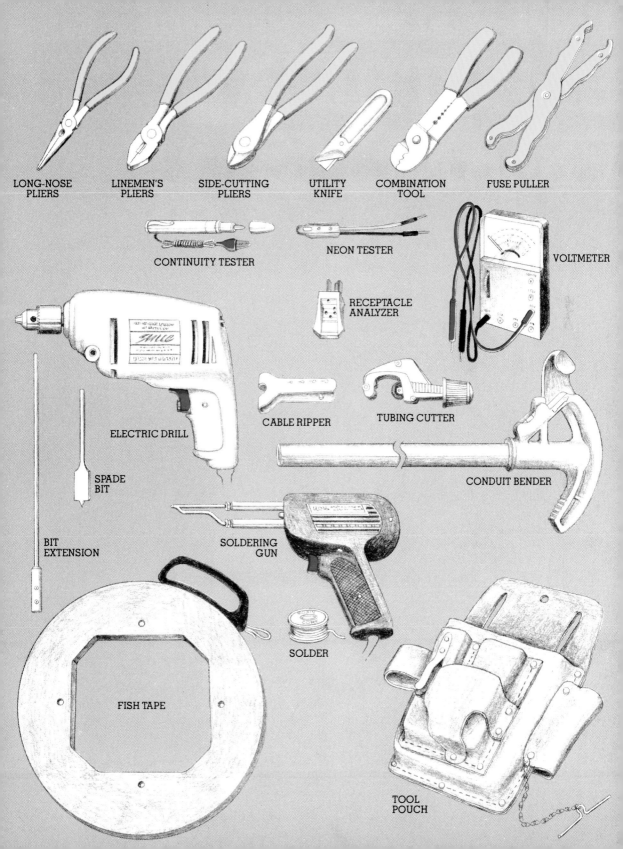

LONG-NOSE
PLIERS

LINEMEN'S
PLIERS

SIDE-CUTTING
PLIERS

UTILITY
KNIFE

COMBINATION
TOOL

FUSE PULLER

CONTINUITY TESTER

NEON TESTER

VOLTMETER

RECEPTACLE
ANALYZER

ELECTRIC DRILL

CABLE RIPPER

TUBING CUTTER

CONDUIT BENDER

SPADE
BIT

BIT
EXTENSION

SOLDERING
GUN

SOLDER

FISH TAPE

TOOL
POUCH

SOLVING ELECTRICAL PROBLEMS

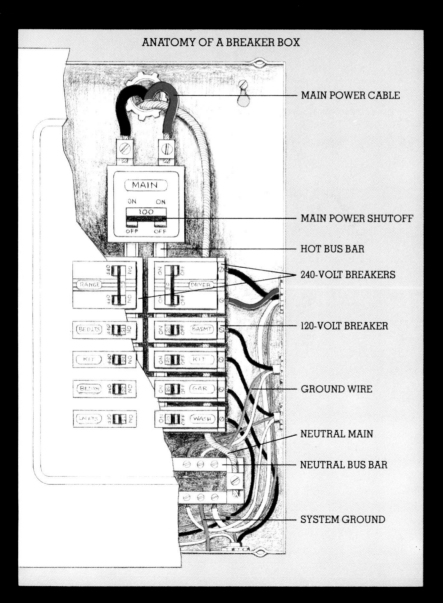

ANATOMY OF A BREAKER BOX

MAIN POWER CABLE

MAIN POWER SHUTOFF

HOT BUS BAR

240-VOLT BREAKERS

120-VOLT BREAKER

GROUND WIRE

NEUTRAL MAIN

NEUTRAL BUS BAR

SYSTEM GROUND

Most electrical projects begin at a service panel such as the breaker box shown at left, or the fuse box on page 14. When a short or an overload shuts down power to a circuit, this is where you go to restore the flow. It's also where you cut off power to a circuit *before* beginning a project.

Power from the meter arrives, in this instance, via those big black and red *main power cables* at the top. They connect to the *main power shutoff.* Turning off this breaker *does not* de-energize the main power cables, but everything else in the panel will be dead.

Note, too, that the main shutoff and the breakers labeled *range* and *dryer* are bigger than the others, and have two hot wires connected to them. These handle 240-volt circuits.

Single hot wire, 120-volt breakers protect individual branch circuits. Each of these devices snaps onto a *hot bus bar,* then is connected to its circuit by a black wire. White wires for each circuit and a bare copper *ground wire* tie into the *neutral bus bar.* Also connected to the neutral bus bar are a *neutral main* and a *system ground,* for reasons explained on page 9.

Troubleshooting Tripped Circuit Breakers

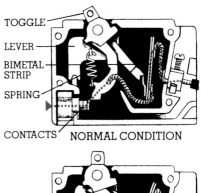

TOGGLE
LEVER
BIMETAL STRIP
SPRING
CONTACTS **NORMAL CONDITION**

spring relaxes

contact is broken

TRIPPED CONDITION

Think of a circuit breaker as a heat-sensing, spring-loaded switch. The cutaway drawings at right show one of these sensitive devices in its normal and tripped states.

When the *toggle* is in its "on" position, current flows through a set of *contacts* held together by a *spring* and *lever*. These are kept in tension by a *bimetal strip*, which is part of the circuit's current flow.

Typically, if something goes awry in the circuit—either a short or an overload—the bimetal heats up, bends, and the spring separates the contacts.

Once current stops, the bimetal cools and tries to straighten again, but it's not strong enough to stretch the spring, so the contacts remain open until the toggle is manually reset.

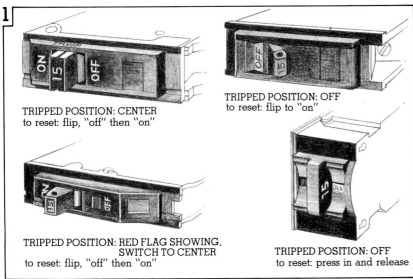

1

TRIPPED POSITION: CENTER
to reset: flip, "off" then "on"

TRIPPED POSITION: OFF
to reset: flip to "on"

TRIPPED POSITION: RED FLAG SHOWING, SWITCH TO CENTER
to reset: flip, "off" then "on"

TRIPPED POSITION: OFF
to reset: press in and release

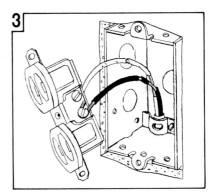

2

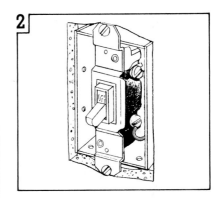

3

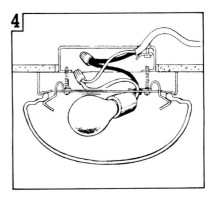

4

1 A tripped breaker may identify itself in any of the four ways shown above. To find out whether the problem that caused the outage has corrected itself, reset the breaker. Don't worry about shock or fire. If there's still a short or overload, the breaker will snap itself off again.

Overloads are easy to relieve. Usually you only need to unplug or turn off one of the circuit's bigger electricity users.

If that doesn't help, suspect a short. To locate one, systematically unplug items until the breaker holds. A defective plug, cord, or lamp socket may be the culprit (see pages 16-19).

2 Short circuits can occur in outlet boxes, too. Here, a wire has pulled loose from the switch and shorted out against the box.

3 Frayed or cracked insulation can expose a wire and cause a short. The solution: Wrap it with several layers of electrical tape.

4 High-wattage bulbs in a light fixture can melt insulation and produce a short. Never use bulbs larger than those for which the fixture is rated. More about troubleshooting light fixtures on pages 22-25.

Troubleshooting Blown Fuses

If yours is an older home that hasn't been rewired, chances are its electrical "heart" is a fuse box rather than the breaker box illustrated on page 12. Fuses and circuit breakers serve exactly the same purpose, but instead of tripping as a breaker does, a fuse "blows" when there's too much current in its circuit. Then you must eliminate the short or overload (see page 13), remove the blown fuse, and screw in or plug in a new one.

Refer to the anatomy drawing below and note that once again power comes in via a couple of *main power cables*. (In a house with no 240-volt equipment, there may be only one of these.)

Next, current flows through a main disconnect, in this case a *pullout block* that holds a pair of *cartridge fuses*.

Next in line are a series of *plug fuses* that protect the black "hot" wires of branch circuits. Unscrewing one of these (handle it only by the rim) disconnects its circuit.

1 Peer closely through the window of a plug fuse and you'll see a strip of metal through which current flows. The intense heat of a short or an overload causes this strip to melt, disconnecting the circuit.

Often just looking at a plug fuse can tell you whether a short or an overload caused it to blow. A short circuit explodes the strip, blackening the fuse window. An overload, on the other hand, usually leaves the the window clear.

2 Cartridge fuses, such as those you might find on the back of a main or 240-volt pullout, can't be checked visually. To determine whether one has blown, either install a new one or check the old one with a continuity tester as shown. The bulb will light if the fuse is good.

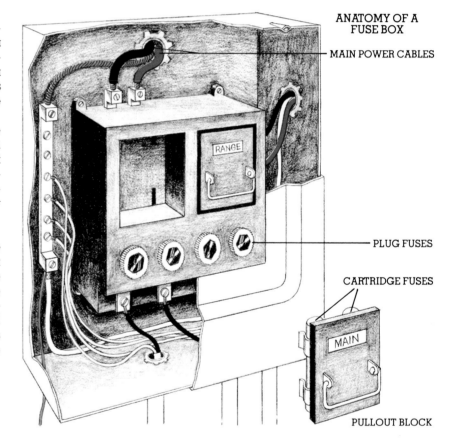

ANATOMY OF A
FUSE BOX

MAIN POWER CABLES

RANGE

PLUG FUSES

CARTRIDGE FUSES

MAIN

PULLOUT BLOCK

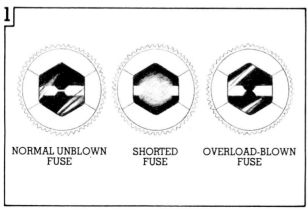

NORMAL UNBLOWN FUSE   SHORTED FUSE   OVERLOAD-BLOWN FUSE

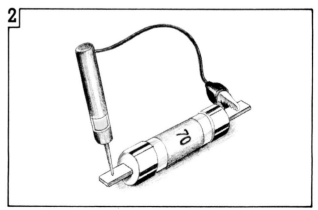

3

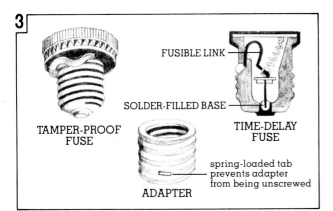

FUSIBLE LINK

SOLDER-FILLED BASE

TAMPER-PROOF
FUSE

TIME-DELAY
FUSE

spring-loaded tab
prevents adapter
from being unscrewed

ADAPTER

4

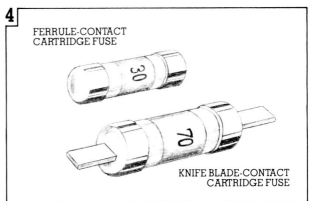

FERRULE-CONTACT
CARTRIDGE FUSE

KNIFE BLADE-CONTACT
CARTRIDGE FUSE

5

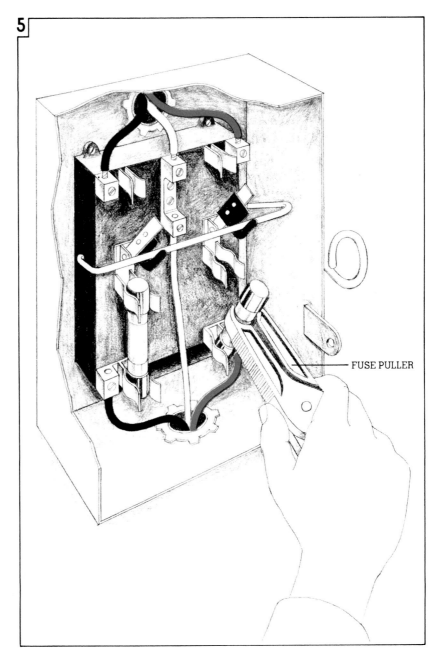

FUSE PULLER

3 Ordinary plug fuses are inter-changeable, regardless of their amperage ratings. So if you have a chronically overloaded circuit, you might be tempted to install a "bigger" fuse. Don't do it. Wiring that gets more current than it was designed to handle heats up and can catch fire.

Tamper-proof fuses make it impossible for you or anyone else to "over-fuse" a circuit. Each comes with a threaded *adapter* that fits permanently into the box. The adapter accepts only a fuse of the proper rating.

Another problem with ordinary plug fuses is that they can be blown by even the momentary overload that often happens when an electric motor starts up. A *time-delay* fuse waits a second or so. If the overload continues, its *solder-filled* base melts and shuts down the circuit. A short blows the *fusible link*, just as it would in any other fuse.

4 Circuits rated at 30 to 60 amps typically use *ferrule-contact* cartridge fuses. *Knife blade-contact* fuses handle 70 amps or more. Both must be handled with extreme caution. Touching either end of a live one could give you a potentially deadly shock.

5 For safety, keep a plastic *fuse puller* with your spare fuses, and use it as shown at right. Note, too, that the ends of a cartridge fuse get hot, so don't touch them even after you've pulled the fuse.

Replacing Plugs, Cords, and Lamp Sockets

Have you ever been zapped by an electrical shock? If so, it probably came from a faulty plug, cord, or lamp socket. These pose the most common shock and fire hazards.

Fortunately, they're also far and away the easiest to repair. Master just a few basics and you can make short work of any potential short circuit.

Plugs get stepped on, bumped against, and yanked out by their cords. Even properly handled ones eventually wear out. So glance at any plug before you use it. If it shows signs of damage, replace it immediately.

The drawing at right shows some of the plug variations you'll encounter. *Round-cord* and *flat-cord* plugs are suitable for lamps, radios, and other low-amperage users. Newer lamp and extension cord plugs are *polarized*, with one blade wider than the other. To learn why, see page 84.

Heaters, irons, and similar appliances pull more current and require heftier plugs; *240-volt* equipment calls for special blade configurations.

Cords vary, too. *Zip cord*, the most common, should be used only for light duty. Twisted cords have two layers of insulation. *Heater cords* are triple-insulated, as is the *heavy-duty*, three-wire type required to ground appliances and other equipment. The *240-volt cords* have three fatter wires.

Lamp sockets come with a variety of different switching mechanisms. Most have brass-plated shells.

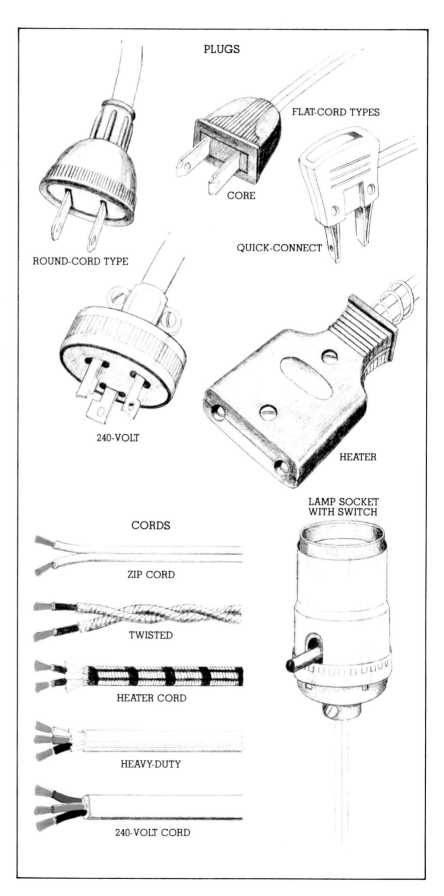

PLUGS

FLAT-CORD TYPES

CORE

QUICK-CONNECT

ROUND-CORD TYPE

240-VOLT

HEATER

LAMP SOCKET WITH SWITCH

CORDS

ZIP CORD

TWISTED

HEATER CORD

HEAVY-DUTY

240-VOLT CORD

Replacing Plugs

Round-Cord Plugs

1 Begin by snipping off the old plug, slipping a new one onto the cord, and stripping away insulation as shown. (For stripping techniques, see pages 86-87.)

2 Now tie the "Underwriters' knot" so that tugging the cord can't loosen the electrical connections you'll be making. Always try to protect connections from stress.

3 Twist the wire strands tight with your fingers. Then, with a pair of long-nose pliers, shape clockwise hooks like these.

4 As you tighten the screws, tuck in any stray strands. If the screws differ, attach the black or ribbed wire to the brass one.

5 Finally, check to be sure all wires and strands are neatly inside the plug's shell, then slip on the cardboard cover.

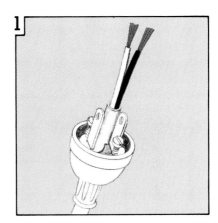

UNDERWRITERS' KNOT
(pull both ends tight)

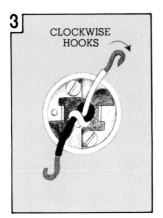

CLOCKWISE HOOKS

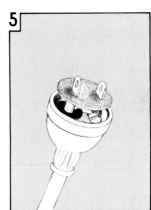

Flat-Cord Plugs

Flat-cord plugs connect in a variety of ways. Some require the same procedure shown above. Others have a core that you attach the wires to, then snap into a shell, as depicted here. Still another core type bites into unstripped cord, such as the quick-connect plug on page 18 does.

1 First slip the shell onto the cord, peel apart the wires, and strip away about half an inch of insulation. (More about stripping wires on pages 86-87.)

2 Twist the strands and form clockwise hooks big enough to wrap three-fourths of the way around each screw.

3 Slip the wires under the plug's screws, then tighten the screws to secure the connection.

4 Now just snap the core into the shell. With practice, the entire job takes only minutes. *(continued)*

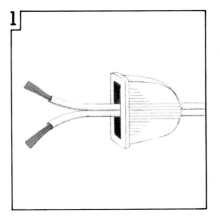

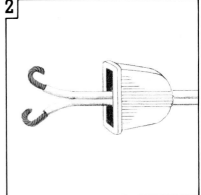

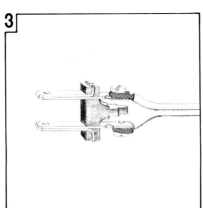

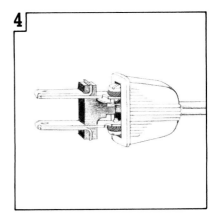

Replacing Plugs (continued)

Quick-Connect Plugs

Keep a few quick-connect plugs on hand and you'll never again be tempted to put off replacing a faulty or questionable plug. Installing one takes about as long as changing a light bulb, and the only tool you need is a sharp knife or pair of scissors.

1 Snip off the old plug. Lift the lever on top of the new plug and insert the zip cord at the side.

2 Closing the lever pierces and holds the wire. It's that easy.

240-Volt Plugs

Really just an over-sized round-wire plug with an extra blade, a 240-volt plug installs in much the same way. You don't have to tie an Underwriters' knot, however. A steel clamp grips the cord.

1 Slide the plug onto the cord and strip the three wires as shown. Twist the strands tightly together and form hooks.

2 Attach the black and red wires to brass-colored terminals, and the green one to the silver-colored screw. Tuck everything into place, tighten the cord clamp, and slip on the cardboard cover.

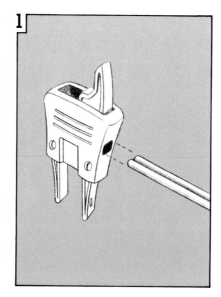
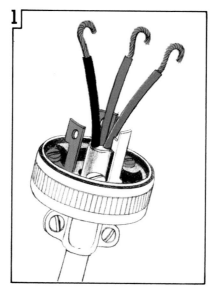

Heater Plugs

Wall plugs for heaters, irons, and other "hot" appliances typically are molded to the cord. To replace the plug, it's best to replace the cord as well. You may or may not be able to replace the female plug at the cord's other end, depending on whether it comes apart.

1 If your plug is like the one shown at right, it will pull apart. Note how the spring relieves strain on the cord.

2 If the plug has a brass terminal, connect the black wire to it, the white to the silver one. Clips tie the plug to the heater blades.

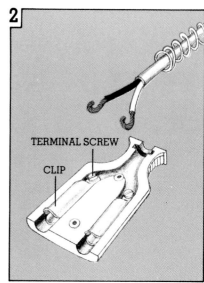
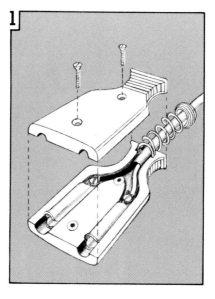

TERMINAL SCREW

CLIP

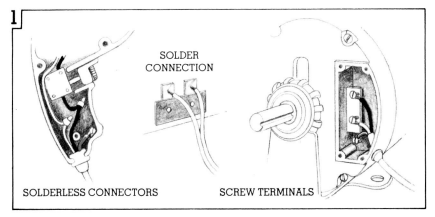

SOLDER CONNECTION

SOLDERLESS CONNECTORS   SCREW TERMINALS

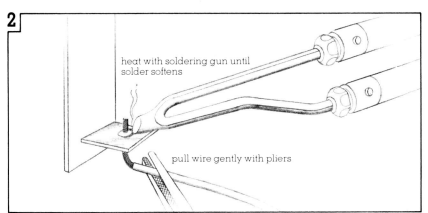

heat with soldering gun until solder softens

pull wire gently with pliers

Replacing Cords

Lamp and appliance cord must be sized according to the load it will carry. For lamps, clocks, and other items that draw less than 7 amps, use No. 18 wire; 7-to 10-amp appliances need No. 16 wire; anything larger than 10 amps should have a No. 14 cord. The larger the wire number, the smaller its size.

1 You'll probably have to dismantle part of the unit to find out how its cord hooks in. Expect to find screw terminals, solder connections, or solderless connectors that screw on or crimp on.

2 Release a soldered connection as shown here. To resolder, insert the wire, heat the entire connection, including the terminal, then touch solder to it. The solder will melt and fuse the joint.

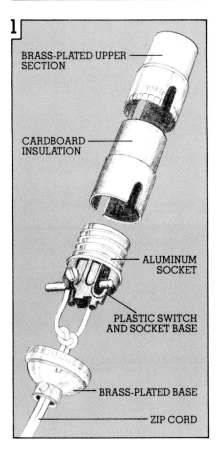

BRASS-PLATED UPPER SECTION

CARDBOARD INSULATION

ALUMINUM SOCKET

PLASTIC SWITCH AND SOCKET BASE

BRASS-PLATED BASE

ZIP CORD

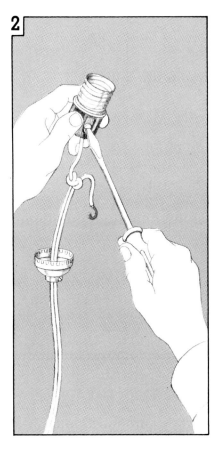

Replacing Lamp Sockets

When a lamp won't work and you know its bulb is OK, unplug the cord and pry up the little brass contact in the socket base. If this doesn't bring results, or if the cardboard insulation has deteriorated, replace the socket.

While you're at it, you might want to install a new cord and plug (a polarized one), too. The easiest way to thread new cord through a lamp base is to tie it to the old one with a piece of string. As you withdraw the old, the new will follow it.

1 Examine the socket shell and you'll find the word *press*. Push hard here and the unit will pull apart into the series of components illustrated.

2 Slip the new socket base onto the cord, tie it with an Underwriters' knot, and attach wires to each of the terminals. Reassemble everything and the job is done.

Replacing Switches and Receptacles

Tired of pampering a balky switch or a paint-glopped receptacle that holds plugs with only the feeblest of grips? Armed with just a screwdriver, a neon test light, and the know-how explained on these two pages, you can install a new one in 15 minutes or less, and that includes time for a couple of trips to your service panel to cut and restore power. That's because all you have to do is wire the new device the way the old one was wired. To learn about some of your choices in new switches and receptacles, see pages 32, 33, 84, and 85.

To be safe, *always* de-energize a switch or receptacle before you touch its inner workings. To do this, you'll need to shut off your home's main circuit breaker or pull its main disconnect fuse block (see pages 12-15 for how to do this). Or deactivate only the circuit you think you'll be working on, then check the condition of the circuit with a test light, as shown here. Note, however, that to test a switch, you must have a good bulb in the fixture it controls. For more information about testing switches, turn to page 23.

Switches

1 Is the circuit live? With the switch off, touch the tester's probes to its screw terminals. If the light glows, the circuit is still hot.

2 If the tester doesn't light, remove the screws holding the switch's ears to the box and pull out the device. Now loosen the screw terminals and disconnect the wires.

3 A three-way switch has three terminals. Note which is marked or otherwise identified as the *common* before you unhook the wire. This terminal may not be in the same place on your new switch.

4 Some new switches have push-in terminals in back, as well as screws on the sides. If you use screws, wrap wires around them in a clockwise direction.

5 Now tuck the wires and switch back into the box and tighten the hold-down screws. Don't force anything; switches crack easily.

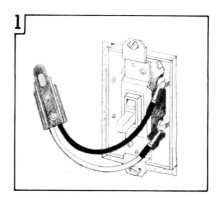

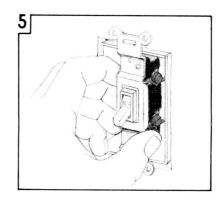

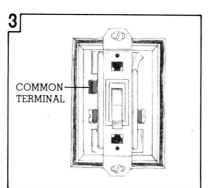

COMMON TERMINAL

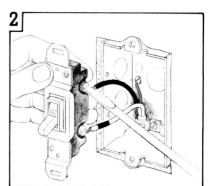

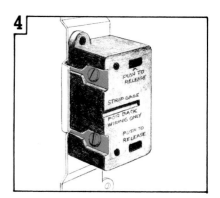

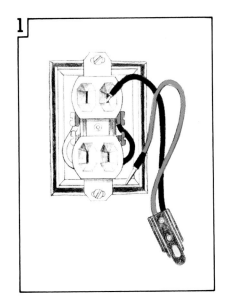

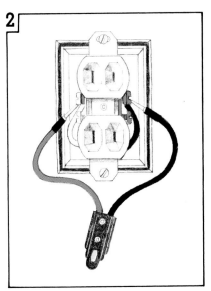

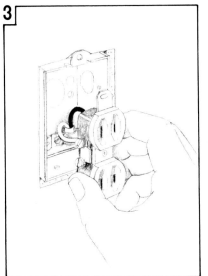

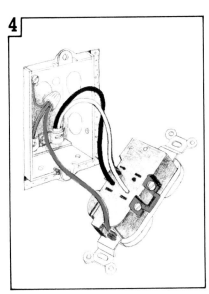

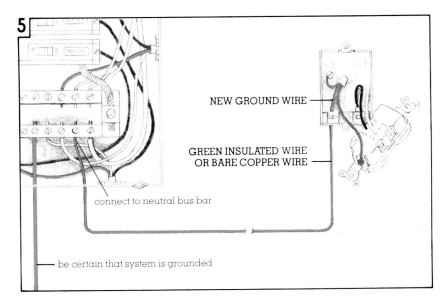

NEW GROUND WIRE

GREEN INSULATED WIRE
OR BARE COPPER WIRE

connect to neutral bus bar

be certain that system is grounded

Receptacles

1 Can you replace a two-slot, non-grounding receptacle with a safer, three-hole version? To find out, remove the wall plate, leave power on, and touch one probe of the tester to the receptacle strap or the box. Now insert the other probe into each of the slots. If either lights up the tester, the box itself is grounded and you can install a three-hole type. If not, either get a two-slot receptacle or — better yet — run a separate ground as shown in the bottom sketch on this page.

2 To learn whether a receptacle is live, touch the tester probes to screws on either side. The light will glow if there's power.

3 Remove the hold-down screws and pull out the device. Be sure to note which wires are attached to which terminals before unfastening them.

4 Newer receptacles have push-in terminals as well as screws. Whichever you use, connect white wires to silver, black to brass. If you find a bare wire inside the box, ground the receptacle's green screw to it and to the box with a couple of short green wires. If there is no bare wire but you know the box is grounded, run a jumper directly from the receptacle to the box.

5 To ground an ungrounded box, run a wire from it to the service panel or to a cold water pipe. Turn to pages 73-76 for information on how to go about fishing wires through finished space.

Troubleshooting Incandescent Fixtures

Incandescent fixtures vary widely in style, but most have some arrangement of the components illustrated in the anatomy drawing at right.

A *canopy plate* attaches to a wall or ceiling fixture box, and also supports a *bulb holder* that consists of one or more *sockets*. *Leads* connect the sockets to wiring in the box.

To get at the bulbs, you usually must remove some sort of translucent glass or plastic *diffuser*. Most manufacturers post a maximum wattage on the canopy. Bulbs of a wattage higher than the recommended rating generate too much heat, which is the main enemy of incandescent fixtures.

When a fixture shorts out, you can almost be certain that the problem lies in the fixture itself or in its electrical box. If one refuses to light, however, the switch that controls it also could be faulty.

(NOTE: Be sure to shut off power before beginning.)

1 Inspect the socket. Cracks, scorching, or melting mean it should be replaced. You can find a new one at a lamp parts store.

If the socket is intact and securely mounted to the canopy plate, remove the bulb and check the contact at the socket's base. If there's corrosion, turn off the circuit — not just the switch — and scrape the contact with a screwdriver or steel wool. Also pry up the contact a bit.

2 If the problem remains, shut off the circuit again and drop the fixture from its outlet box. To do this, remove either a single nut in the center or a pair of screws located off-center. Now check for loose connections and for cut, frayed, or melted insulation. Wrap any bare wires with electrical tape.

ANATOMY OF AN INCANDESCENT CEILING FIXTURE

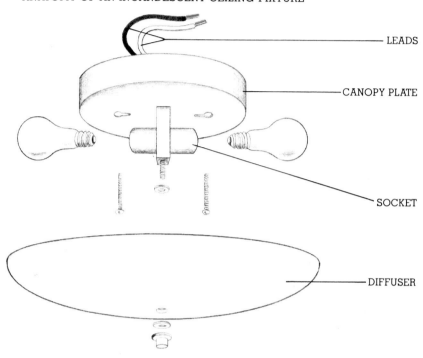

LEADS

CANOPY PLATE

SOCKET

DIFFUSER

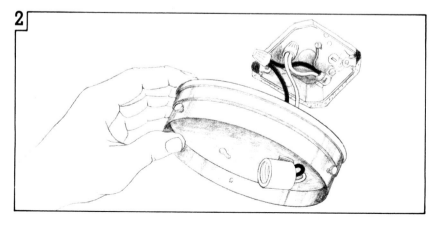

Testing Switches

1 Internal switches on a fixture or lamp usually are connected with a pair of small pressure connectors. **To test the switch, unplug the lamp and remove the connectors holding the switch's leads.** Leave the bare wires twisted together but arrange them so the connections aren't touching each other or anything else.

Restore power to the lamp or fixture and carefully touch your tester to the connections. Now flip the switch and test again. If the tester lights, the switch is bad.

2 To test a wall switch, turn it to the "on" position, and touch the tester to the terminal screws. Here, too, if the tester lights, the switch must be replaced, as shown on page 20.

3 Nervous about poking into a live switch? Then shut down its circuit and conduct your investigation with a continuity tester. In the above situations, the tester should light when the switch is on, but not when it's off. If you get a glow in both positions or no glow in either position, you need a new switch.

To check out a three-way switch such as this one, shut off the circuit and attach the tester's clip to the common terminal; it's usually identified on the switch body. Now touch the probe to one of the other terminals and flip the switch. If it's OK, the tester will light in one position or the other. Repeat this test with the other terminal.

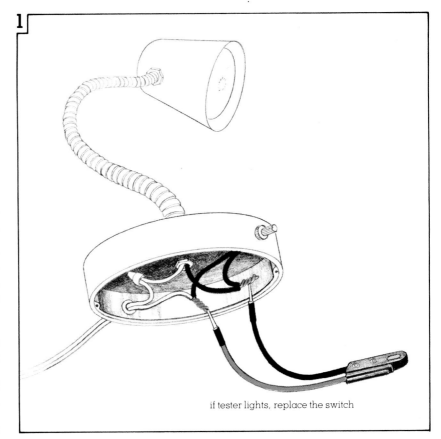

if tester lights, replace the switch

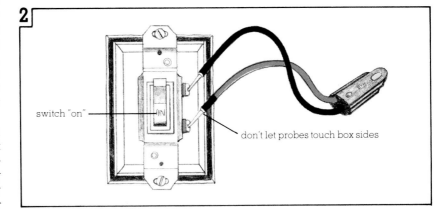

switch "on"

don't let probes touch box sides

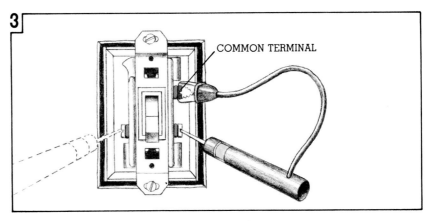

COMMON TERMINAL

Troubleshooting Fluorescent Fixtures

Switching on an ordinary light bulb charges a metal filament that literally burns with white heat. Fluorescent tubes, on the other hand, don't get nearly as hot because they operate in a different way.

Anatomically (see sketch below), the heart of a fluores-cent fixture is its *ballast*, an electrical transformer that steps up voltage and then sends it to a pair of *lamp holders*. The current from the lamp holders excites a gas inside the tube, causing its phosphorus-coated inner surface to glow with cool, diffused light.

Because they produce far less heat, fluorescent tubes last much longer than incandescent bulbs and consume considerably less electrical energy. Problems are fewer, too. Following are the major ills you're likely to encounter with fluorescent fixtures.

ANATOMY OF A RAPID-START FLUORESCENT LIGHT

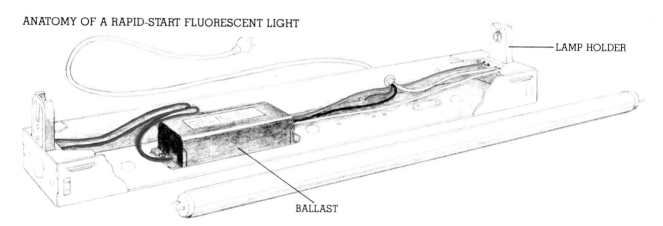

LAMP HOLDER

BALLAST

1 Rarely do fluorescent tubes burn out abruptly. When a tube won't light, try wiggling its ends to be sure they're properly seated.

If this doesn't get things glowing again and yours is a rapid-start tube like the one shown, suspect a loose or broken connection. **Start your search by turning off power to the circuit,** removing the tubes and the fixture cover, and inspect-ing all connections inside. Next, check the switch (see page 23).

2 As a last resort, you may have to drop the entire fixture and look for loose connections and broken or bare wires in the outlet box.

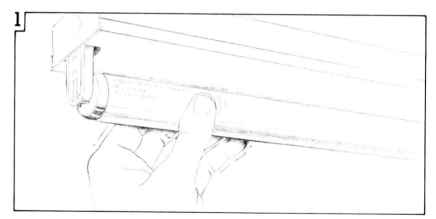

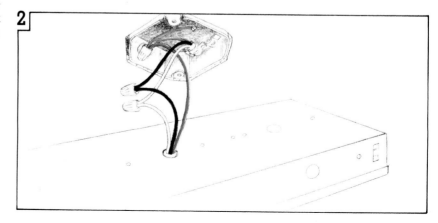

3 When a tube begins to fail, the normally grayish bands near its ends gradually blacken. Uniform dimming usually means the tube simply needs washing. When you shop for a new tube, select one of the same wattage as the old one.

4 Older, delayed-start fluorescent lights flicker momentarily as they light up. If the flickering continues, make sure the starter is seated by pushing it in and turning clockwise.

When the ends of a tube light up but its center does not, the starter probably has gone bad. To remove it, press in and turn counterclockwise.

5 Humming, an acrid odor, or tar-like goop dripping from a fixture indicates that the ballast is going bad. To replace it, follow the procedures shown below. But first compare the price of a new ballast with that of a new fixture. You may be better off replacing the entire unit.

3
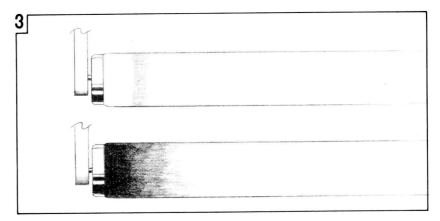

4
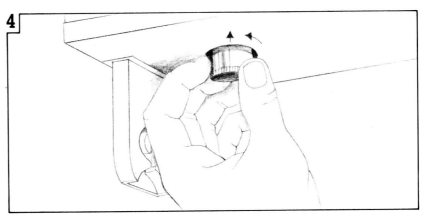

5
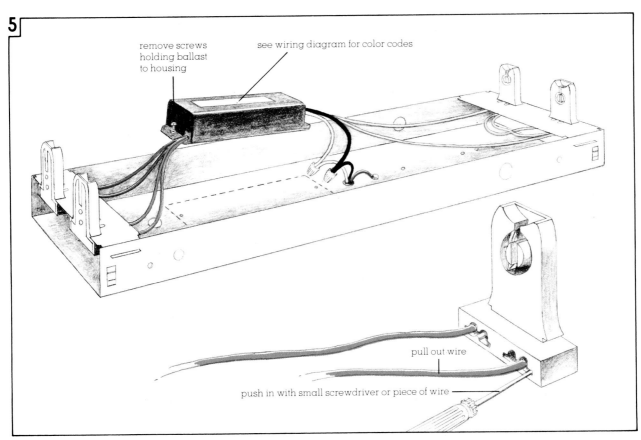

remove screws holding ballast to housing

see wiring diagram for color codes

pull out wire

push in with small screwdriver or piece of wire

Repairing a Silent Door Chime

Apprehensive about electricity's potential for fire and shock? You can put aside those fears when you work with door chimes, intercoms, thermostats, and even some lighting systems. Current for them flows in the same sort of circuits illustrated early in this book, but at voltages that scarcely tickle.

Examine the chime systems shown at right and you'll see that everything starts with a *transformer*, which "steps down" 120-volt house current to one of several levels between six and 30 volts. From the transformer, light-gauge wire makes a circuit that's normally open. Pressing the button, which is a spring-loaded switch, closes the circuit, sending the lowered voltage to sound the chimes.

Troubleshooting low-voltage circuitry is a simple process of elimination. A voltmeter makes the job easier, but you can also do most tests with only a short length of wire or a screwdriver.

1 Weather and insistent deliverymen make buttons the most vulnerable parts of a chime system, so start your investigation here. To test a button, unscrew it and jump its terminals as shown. (You may have to scrape away corrosion first.) If the chimes sound, the button is faulty and should be replaced.

If the chimes don't sound, the problem is elsewhere. Disconnecting the button and twisting its wires together let you make the other tests shown on the opposite page without running back and forth to push the button.

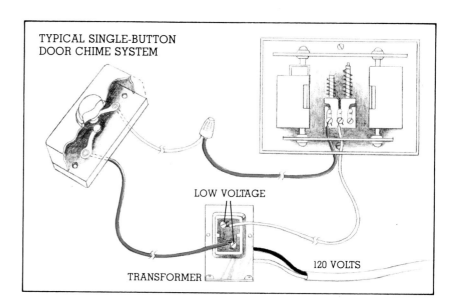

TYPICAL SINGLE-BUTTON DOOR CHIME SYSTEM

LOW VOLTAGE

TRANSFORMER

120 VOLTS

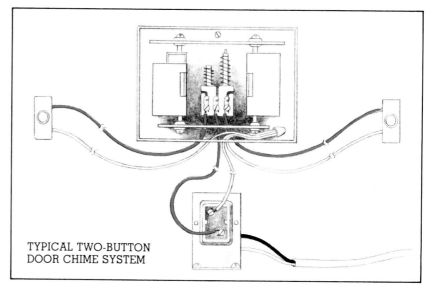

TYPICAL TWO-BUTTON DOOR CHIME SYSTEM

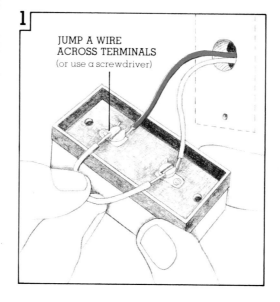

1

JUMP A WIRE ACROSS TERMINALS
(or use a screwdriver)

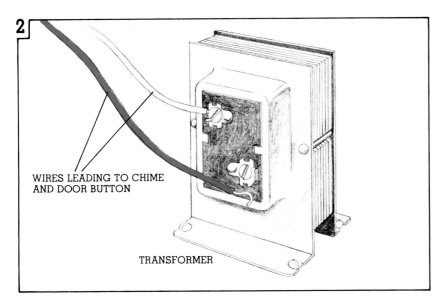

2

WIRES LEADING TO CHIME
AND DOOR BUTTON

TRANSFORMER

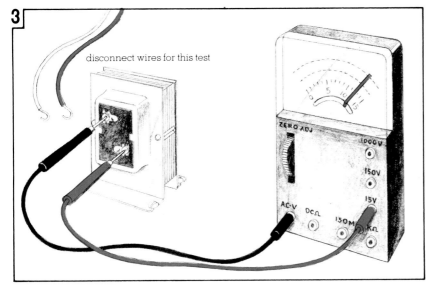

3

disconnect wires for this test

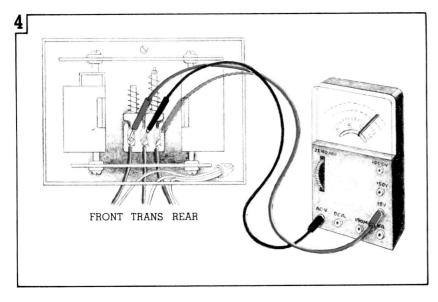

4

FRONT   TRANS   REAR

2 Next, take a look at the transformer. It may be situated at or near your home's service panel, or attached to a junction box somewhere in the vicinity of a door or the chimes themselves.

Here, one of the wires has come loose. Reconnecting it should ring the chimes.

3 Is the transformer working? To find out, disconnect both wires and touch each terminal with the probes of a voltmeter. If the meter shows no reading at all, the transformer is the culprit and must be replaced.

You also can test a transformer by jumping its low-voltage terminals with a screwdriver. If you see even a weak spark, the unit is OK.

Before disconnecting a dead transformer, be sure to shut off the household circuit it taps into. Here you're dealing with 120 volts, and all the usual cautions apply.

4 If the transformer passes your tests, examine the chime unit itself. First look for loose or broken connections.

Now, to check the front chimes of a two-button system, touch the voltmeter probes across the *front* and *trans* terminals. If the meter registers a reading, the chimes are shot. To test the rear chimes, touch the *rear* and *trans* terminals.

If you don't have a voltmeter, test the chimes by connecting them directly to the transformer. Good ones will ring.

If the button, transformer, and chimes all check out, a wire has probably broken somewhere. To confirm this disconnect the chime's leads and test for current flow with the voltmeter, or jump them with a screwdriver and watch for a spark. To learn about running low-voltage wiring, see page 54.

MAKING ELECTRICAL IMPROVEMENTS

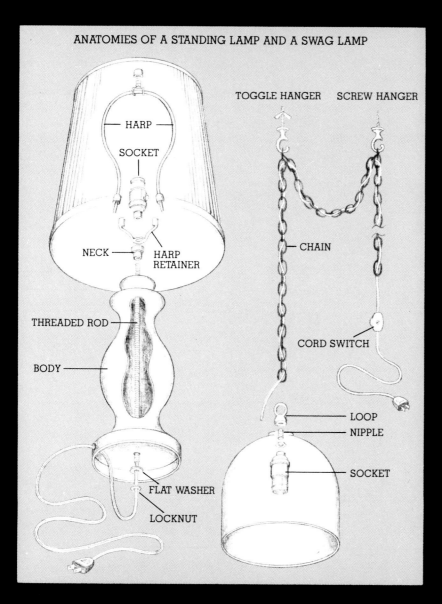

ANATOMIES OF A STANDING LAMP AND A SWAG LAMP

HARP

SOCKET

NECK

HARP RETAINER

THREADED ROD

BODY

FLAT WASHER

LOCKNUT

TOGGLE HANGER   SCREW HANGER

CHAIN

CORD SWITCH

LOOP
NIPPLE

SOCKET

The previous section tells how to cope with electrical nuisances that afflict any household from time to time. Now for some fun: a dozen projects that can help you live better electrically.

Our first may be the most creative, and it's certainly the easiest. With just a handful of components such as the ones illustrated here, you can craft your own standing lamp or swag lamp.

Drop by a lamp store and you'll find parts that can answer almost any need. Most standing lamps consist of a *body*, sometimes a *base*, and a *harp* to support the shade. Working elements include the same *plug*, *cord*, and *socket* setup you learned about on pages 16-19. Inside the lamp, the cord usually runs through *threaded rod*, which you can buy in standard lengths and cut to size.

With a swag lamp, the cord threads through a *chain*, which attaches to a *loop*. This loop is connected to a socket with a *nipple*. Special *toggle* or *screw hangers* and *hooks* hold the chain, and a *cord switch* or *pull switch* controls the light.

Wiring Your Own Lamps

Standing Lamps

A handsome hunk of wood or marble, an old-fashioned kerosene lamp, a favorite vase —just about anything hollow or with a hole through its center can be fashioned into a lamp that stands on a table or the floor. The monkey shown below came from a ceramics shop that specializes in lamp bodies.

1 To get the monkey on its feet, feed a rod up through it and the base. At the bottom you can secure it with a *locknut* and *washer*, or two ordinary lamp nuts; above, a more decorative *knurled nut* or a threaded *neck* holds everything together. Fit a *harp retainer* over the threaded rod then screw on a *socket base* (most lamp parts have the same ⅜-inch threads), and your main assembly job is done.

2 With shorter bodies, you'll probably be able to simply push cord up through the rod. Taller versions may require running stiff wire through first, then pulling the cord. At the top, tie an *Underwriters' knot* as shown. Secure the cord below with a second knot, if desired.

3 Now strip insulation and attach the cord's leads to the socket terminals. A plug, of course, goes on the other end. To learn more about installing plugs, cords, and sockets, turn to pages 16-19.

4 Finally, reassemble the socket and install the harp as shown. Add a bulb and a shade, and "Mr. Monkey" is ready to light up a child's room. *(continued)*

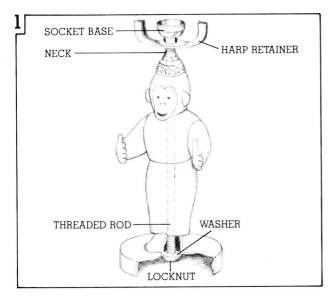

1
SOCKET BASE
NECK
HARP RETAINER
THREADED ROD
WASHER
LOCKNUT

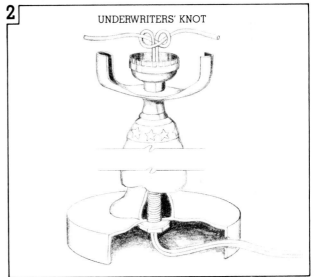

2
UNDERWRITERS' KNOT

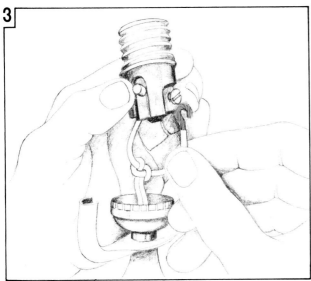

3

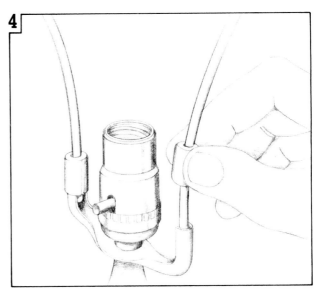

4

Wiring Your Own Lamps *(continued)*

Swag Lamps

Unlike standing lamps, swags derive their support from above. Although the lamps you can create may vary considerably in style, they all go together in much the same way.

1 You can buy lamp chain by the running foot in a variety of finishes. Attach a loop to the chain by spreading open an end link, then crimping it back together.

2 Cord also comes in a spectrum of colors. One version has transparent insulation. When threaded through a chain, it is almost invisible.

3 Next, fasten the loop to the shade support assembly. With a shade that is open at top and bottom, the loop fastens to it via a three-pronged *spider*, as shown here. Screwing on the socket base secures the connection.

    With another popular type of swag, a *shade holder* fits between the loop and socket.

4 Tie the Underwriters' knot as shown, strip about ¾ inch of insulation from each of the cord's wires, and secure the cord to the socket body.

5 If there's a ceiling joist near where you want to hang the lamp, bore a hole and install a *screw hanger*. No joist? Use a *toggle hanger*. You'll also need another hook or two to get the cord and chain to a wall receptacle. Fit the cord with a plug and switch.

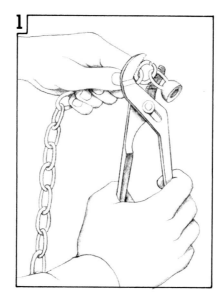

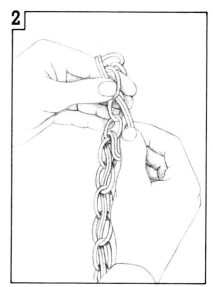

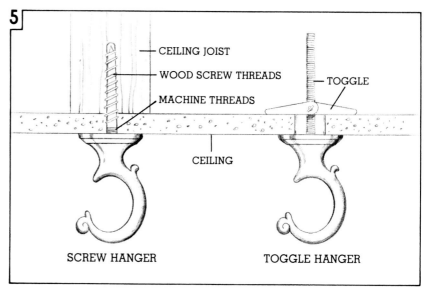

NIPPLE

SPIDER

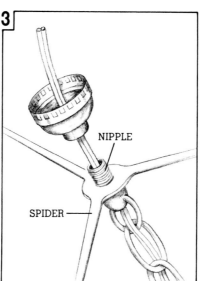

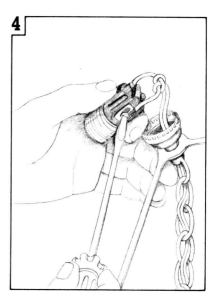

CEILING JOIST

WOOD SCREW THREADS

TOGGLE

MACHINE THREADS

CEILING

SCREW HANGER

TOGGLE HANGER

Hanging New Fixtures and Track Lights

Looking for a quick, relatively inexpensive way to redecorate a room? A new fixture will cast an entirely different light on things. Your choices are legion, ranging from simple, up-against-the-ceiling bubbles to elaborate track layouts, but almost all of them mount as shown in the drawings below.

1 **Shut off the power to the old fixture, then disassemble it.** To drop the fixture, remove the screws or locknut holding the canopy in place. Disconnect the black and white *leads*.

2 Instructions with your new fixture will show how to mount it. With many track light setups, a power module connects directly to the house wiring. The track, connector fittings, and the lights themselves become electrified as soon as they are hooked into the module. With another type, shown here, power comes directly into the end of a length of track via a pair of leads.

3 This hanging fixture simply attaches with a pair of *studs* screwed to the box. You connect the black and white leads (and the stranded ground wire), coil them up into the box, and secure the canopy with *cap nuts*.

4 Alternative mounting systems include a *strap*, which you'll need if holes in the canopy don't mate with those in the box. A strap also lets you install a *nipple* and *locknut* for center mounting. Or use a *hickey* screwed to a *stud*.

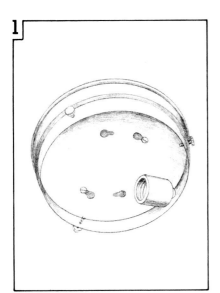

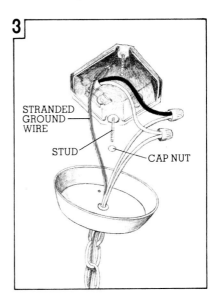

STRANDED
GROUND
WIRE

STUD

CAP NUT

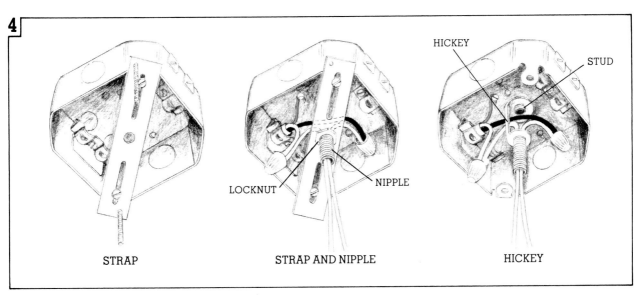

STRAP

LOCKNUT   NIPPLE

STRAP AND NIPPLE

HICKEY

STUD

HICKEY

Installing Dimmer Switches

When it comes to decorating with a light touch, a dimmer switch gives you the ultimate in mood-setting versatility. Lower the glow to candlelight level or rev up to full power. A dimmer puts control at your fingertips.

Dimmers are thrifty, too. Operating a bulb at less than its full intensity saves energy and prolongs the life of the bulb. If you haven't shopped for a dimmer recently, you may be surprised to learn that now there are alternatives to the "push/push" rotary switch most of us are familiar with.

The one shown in the first illustration below, for instance, is touch-sensitive, like a modern elevator button. Touching its flush plate turns lights on and off; holding a finger on it adjusts the light level.

No matter what type of dimmer you select, make sure you know its limitations. Ordinary dimmers, for example, can only handle up to 600 watts. For higher wattage situations, you'll have to buy more expensive units of 800, 1,000, or 1,500 watts.

Realize, too, that fluorescent lights require special dimmers and ballasts, as explained on the opposite page.

(NOTE: Always shut off power to the circuit before beginning your installation.)

Incandescent Dimmers

1 Some dimmers have ordinary screw terminals, others have a set of *leads*. Hook up a single-pole lead type as shown here. For terminal connections, see page 20.

Most dimmers are deeper than conventional switches, which means you may have to re-arrange wires in the box before you can fit one in. Don't force dimmers because they crack easily. For crowded situations you can order thin-profile units.

2 Three-way dimmers have three hot leads. Before you remove the old switch, determine which is its common terminal and hook that wire to the dimmer's *common lead* (see page 20). With most three-way dimmers, you install only one per situation. The second switch must be a three-way toggle. One exception is the three-way version of the touch dimmer shown in drawing 1. Install these only in pairs.

3 Some cord dimmers, such as this one, automatically pierce the insulation. With others, you have to strip wires.

4 Socket dimmers hook up like this. They have the same diameter as conventional sockets, but are a little taller.

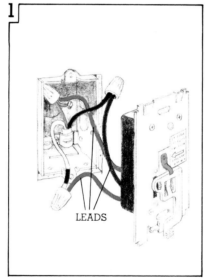

LEADS

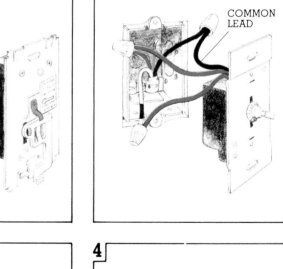

COMMON LEAD

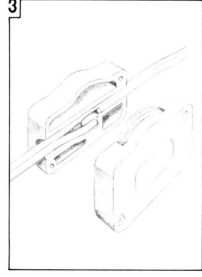

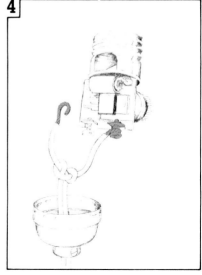

Fluorescent Dimmers

Fluorescent dimmers that will control from one to eight lamp units install in exactly the same way as their incandescent cousins on the opposite page. But you also must equip each of the lamps with a special ballast.

1 Shut off power to the circuit, remove the lens, tube(s), and cover plate, then disconnect the fixture leads from the house wires. Take down the fixture and lay it on a table so you can easily work on it.

2 Next, make a diagram showing which wires connect where. Remove the *lamp holders* and disconnect their wires by poking into the terminals with a thin nail. Remove the old *ballast*, install a new one, and connect it to the lamp holders. Re-install the fixture, turn on the power, and adjust the dimmer as explained in instructions that come with the unit.

3 Here's how to hook up fluorescent fixtures in tandem. Dimming ballasts (remember, you'll need one for each fixture) are available only for four-foot-long, 40-watt, rapid-start lamps. For more about fluorescent lighting, see pages 24 and 25.

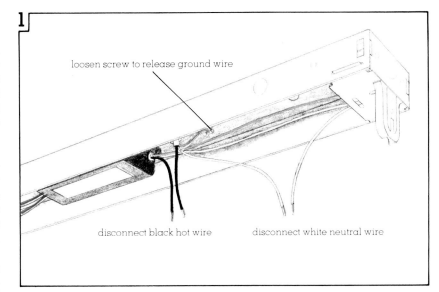

1

loosen screw to release ground wire

disconnect black hot wire   disconnect white neutral wire

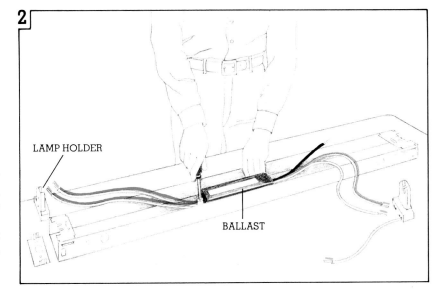

2

LAMP HOLDER

BALLAST

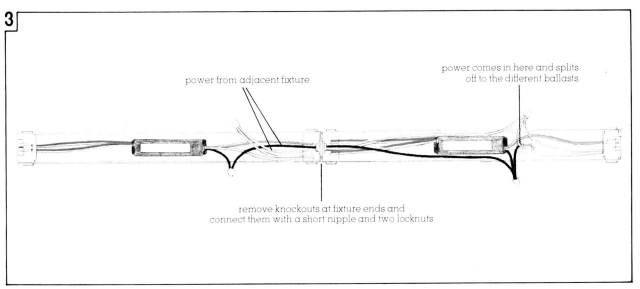

3

power from adjacent fixture

power comes in here and splits off to the different ballasts

remove knockouts at fixture ends and connect them with a short nipple and two locknuts

Installing a Timed-setback Thermostat

A clock-style thermostat that automatically adjusts itself to a more economical level while you sleep or when no one is at home can conserve energy. In most cases, the greater the setback, the more you save.

But be advised that setting back might not make sense if your home has a heat pump. On cold mornings—the chilliest part of the day—the heat pump sometimes won't be able to "catch up" by itself. When it can't, backup electric resistance heating can wipe out savings.

Before shopping for a timed-setback unit, pop the cover plate from your present thermostat and count the number of wires leading to it. Heating-only thermostats usually have two wires; heating-cooling models have four or five wires. The new unit must be compatible.

1 Shut down the power, remove the cover from your old thermostat, and loosen the wires connected to its terminals. Keep track of which wires go to which screws. If your furnace circuit has only two incoming wires, you can use a thermostat with a clock powered by quartz batteries that automatically recharge from current in the circuit.

2 Some thermostats consist of a single, integral unit. Others mount on a separate *wall plate*, like this one. Take care that the wires don't drop into the wall cavity when you remove the thermostat or plate.

3 Before you secure the new base, make sure it's level. If it isn't, the thermostat won't work properly. After the base is mounted, push excess wire back into the

wall and hook up the leads according to the manufacturer's instructions.

4 Finally, mount the body to the base plate. After you've turned the power back on, set the clock, then clip *program pins* at the times you want the setback to begin and end. Here extra pins allow for up to three setbacks in a 24-hour period.

How many degrees should you set back? Finding the optimum spread for your house may call for experimentation. First, try a setting that's 10 degrees lower or higher than your normal heating or cooling levels. If the house cools down or warms up too much, decrease the setback to eight degrees.

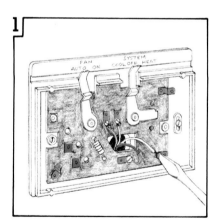

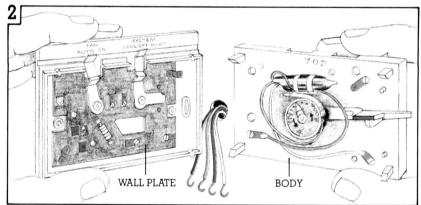

WALL PLATE BODY

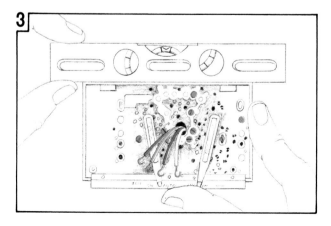

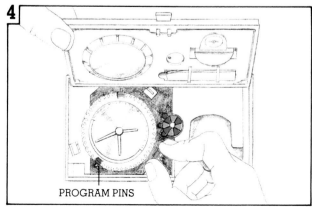

PROGRAM PINS

Wiring Receptacles, Fixture Outlets, and Switches to Existing Circuits

Tired of hassling with a long extension cord or groping around in a dark closet to find something? You can eliminate these and other irritations by installing a new handily located receptacle or a switch-controlled light fixture.

Adding an outlet or two needn't be a big job if you can tap into an under-used circuit nearby. This page shows how to compute whether a circuit can handle additional load, and where you can tap in for power. Then the following pages take

you step by step through nine common new receptacle, light, and switch hookups.

What if you can't find a circuit with capacity to spare? Then you'll have to bring in a fresh one from the service panel, as shown on pages 56-59.

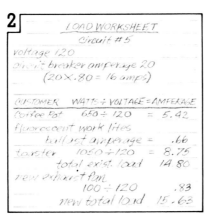

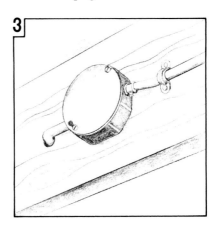

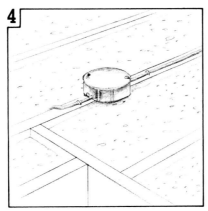

EMPTY
TERMINAL

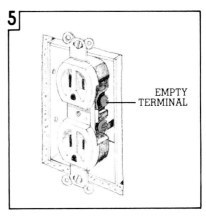

Determining the Load

1 Start by identifying the customers on a circuit. It's a process of elimination. Lights that go out and appliances that won't work when a circuit is shut down are on that circuit. Log your findings on the inside of the service panel cover.

2 Now prepare a worksheet to determine whether the circuit you plan to tap can handle the extra load. If the total amperage of its

regular customers doesn't exceed 80 percent of the circuit's rated capacity, you can add to it.

Where to Tap In

3 If your basement ceiling is open, look for a junction box near where you need the new outlet. To get a cable to it, you may have to "fish" through walls (see pages 73-76).

4 Or check the attic. Here you might find a junction box atop the joists, as shown, or buried under

insulation. Just make sure you're not tapping into the box for a ceiling fixture.

5 Consider nearby receptacles, too. If you find one with a set of unused terminals, you can fish wire from here to the new location. A prime candidate might be in an adjacent room near where you want the new outlet.

A New Approach To Understanding Wiring Diagrams

One reason why ordinary wiring diagrams seem baffling at first is that they show everything already hooked up. To figure out what's going on, you have to trace each wire's path, which is like trying to understand how to tie a knot by untying one.

These drawings use another approach. Follow the sequences and you'll see circuit extensions develop a step at a time.

Adding Another Receptacle By Tapping an Existing One

Let's start with one of the simplest situations: tapping into a duplex receptacle. Remember that this will work only if the receptacle is at the end of a wiring run. If it's not, you'll find all the terminals already occupied.

(NOTE: Be sure to shut off power before beginning.)

1 Remove the screws that secure the existing receptacle to the box. If it's not a grounded type, consider replacing it with one that is (see page 21).

2 Cut an opening for the new box and fish cable from the existing outlet (see pages 73-76).

Once the cable is in place, connect it to the existing and new boxes (see page 77), and install the new box (pages 66-67). Hook up the grounds as shown.

3 Connect the black wire to the hot side of both the new and existing receptacles. Hot terminals have brass-colored screws.

4 Make the white and ground connections to the silver and green screws, and screw the receptacles to their boxes. Finally, turn on the power and test your installation (see page 21).

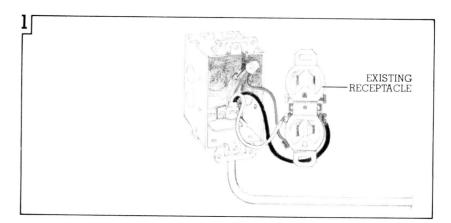

1

EXISTING RECEPTACLE

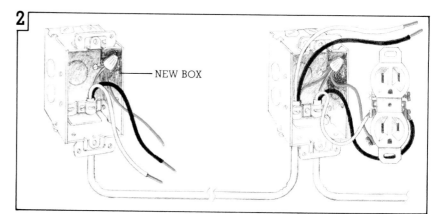

2

NEW BOX

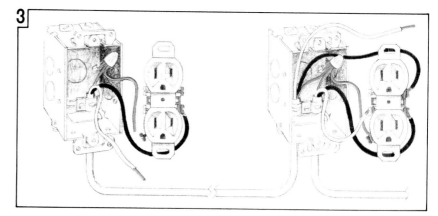

3

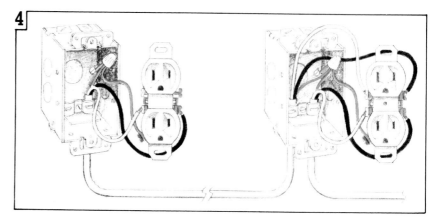

4

Splitting a Receptacle and Controlling It With a Switch

Examine a duplex receptacle and you'll see that the sets of terminals on either side are connected only by a small metal tab. Break this bridge and the upper and lower receptacles can function independently.

This comes in handy in several situations. In a kitchen or shop, for example, you might want to connect a heavily used outlet to two circuits. Or you might want to control half the receptacle with a switch so you can turn a living room lamp on and off from a doorway.

(NOTE: Be sure to shut off power before beginning.)

1 Disconnect the wires from the receptacle, then run two-conductor cable and install a new box, following exactly the same steps illustrated on the opposite page. Because it's almost impossible to fish cable laterally inside walls, you'll probably have to route it to the basement or attic, then bring it up or drop it down from there. Hook up ground wires.

2 Now route the circuit's hot leg to the switch, using the jumper connection shown in the existing receptacle box.

3 Wrap the white wire with black tape to show that it's hot, too, then connect it to the switch and receptacle terminals. Next, connect the remaining black wire to the receptacle as shown. Snap off the brass-colored tab with long-nose pliers to split the outlets, but leave the silver-colored tab in place.

4 Connect the white and ground wires to the receptacle, screw the devices to their boxes, turn on the power, and make your tests. Here the upper outlet will be live only when the switch is on. The lower one remains hot at all times.

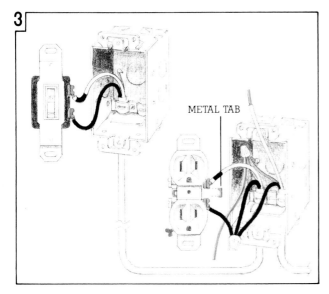

NEW BOX

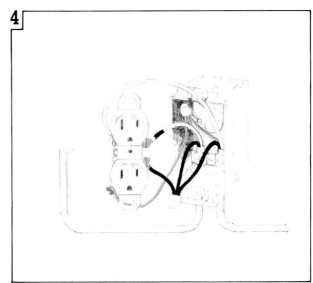

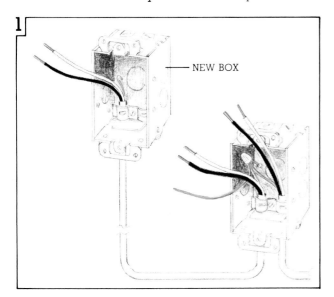

METAL TAB

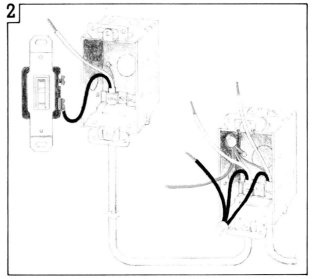

Installing a Ceiling Fixture (With Switch Beyond)

Fixture installations get a bit tricky because you wire them differently depending on where they're located in relation to the power source and the switch or switches that will control them. Will power flow through the switch to the fixture, or vice versa? Here, the switch is beyond.

(NOTE: Be sure to shut off power before beginning.)

1 Open up a junction box and you'll see something like this. You may see even more wires. If you do, locate the circuit you want by testing sets of wires one by one with a continuity tester.

2 Run cable and install the new fixture and switch boxes. Connect the ground wires as shown.

3 Connect the black wires as shown. Note how the black wire picks up power at the junction box and takes it to and through the fixture box and on to the switch.

4 Now mark the white wire running between the fixture and switch boxes with black tape. Hook one end of it to the switch.

5 Finally, connect the black-taped white wire to the black fixture lead, the untaped white wire to the white lead, and all the white wires at the junction box.

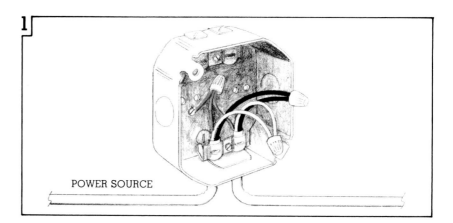

POWER SOURCE

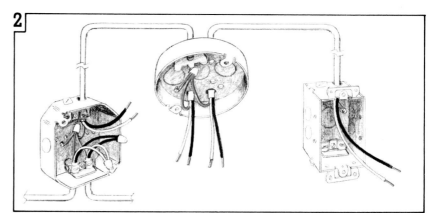

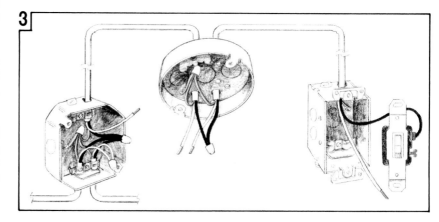

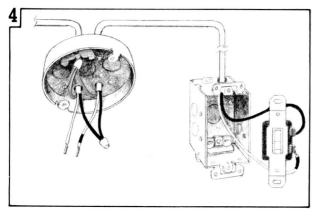

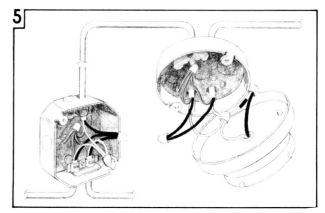

Installing Two Ceiling Fixtures Controlled by One Switch (Fixtures Beyond Switch)

In this installation, power comes to the switch first, then goes on to both of the fixtures. A single switch can control as many fixtures as you like if you extend the run from one to the next. Make sure, though, that the wattage total doesn't exceed the maximum indicated on the body of the switch. Toggle switches are rated in amps; dimmer switches list maximum wattage.

(NOTE: Be sure to shut off power before beginning.)

1 Fish cable, install new boxes where you want the switch and fixtures, and connect the ground wires as shown. Power can come from a junction box or receptacle with an empty set of terminals, as shown on page 36.

2 Hook the black wires to the switch's terminals. These carry current to the fixtures when the switch is on, but not when it's in the off position.

3 At the first fixture you'll have three black wires. Connect them all together as shown. At the second fixture hook its black lead to the incoming black wire.

4 Since the switch interrupts only the circuit's hot leg, you don't connect white wires to it. Attach them to each other and to the fixture's white leads as illustrated.

Note: You can control several fixtures with a switch that's beyond them, too. Just route power as was done on the opposite page, then back to the fixtures. If you want to operate two lights with separate switches, turn the page and we'll show how that's done.

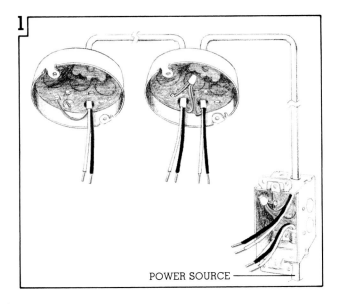

POWER SOURCE ⸺

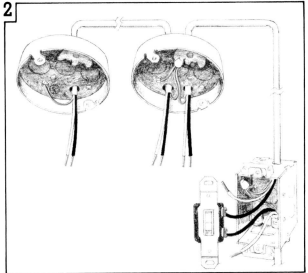

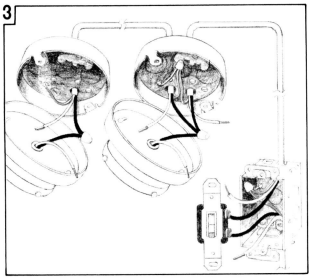

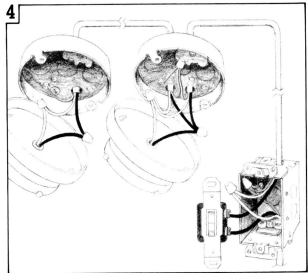

Installing Two Fixtures Controlled By Separate Switches

If you're going to the trouble of installing two ceiling outlets and a wall box, you might be surprised to learn that it's only a little more work to provide individual switches for the fixtures. You simply "gang" together two boxes for the switches and connect them to the fixtures with three-wire cable.

Electricians are fond of three-conductor cable because in most instances it can do twice the work of the two-conductor version. Study this example and you'll be ready for the intricacies of three- and four-way switching that follow.

(NOTE: Be sure to shut off power before beginning.)

1 Here, three-conductor cable goes from the switch box to one fixture, then on to the next. Power comes to the fixtures first via two-conductor cable.

2 Bring power to the switches by connecting the black wires as shown. Short lengths of wire let you attach to switch terminals.

3 Now complete the switch hookup with the red and white wires. Be sure to code the whites with tape to show that they're hot.

4 Complete the circuit by connecting the red and white wires to the fixture leads as shown. If power in your installation will come first to the switches, split the incoming black wire as shown here, and run the outgoing red and black wires from the switches to the fixtures. The neutral wire, which is "shared" by both of the switch circuits, goes right on through, as shown on page 39.

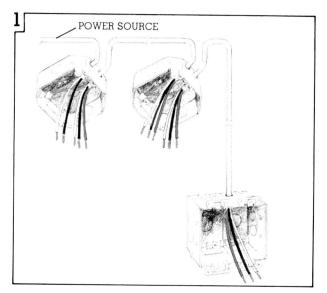

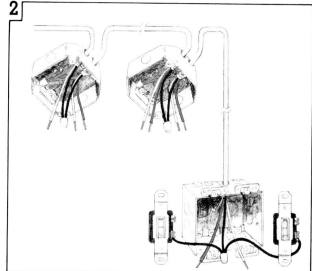

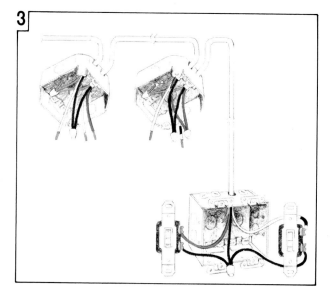

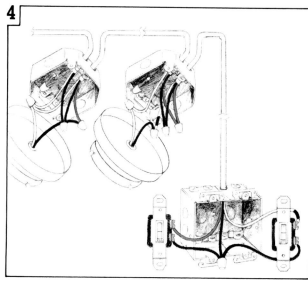

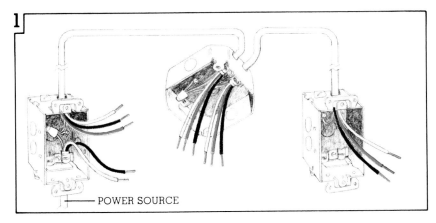

1

POWER SOURCE

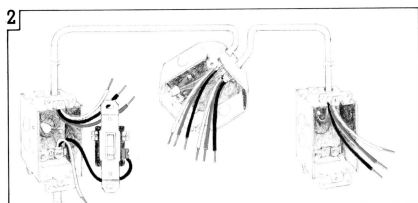

2

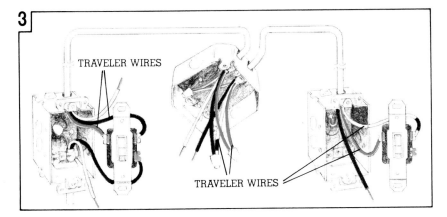

3

TRAVELER WIRES

TRAVELER WIRES

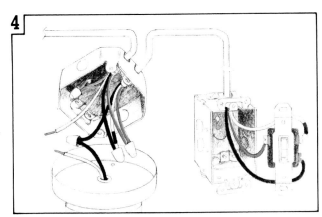

Installing Three-way Switches—Case No. 1

Prepare for numerical confusion when you deal with three-way switches. First of all, they control fixtures from two locations, not three. Secondly, they do have three terminals and need three-conductor circuits. And finally, you hook them up in one of three ways, depending on where the switches are in relation to the power source, the fixture, and each other.

(NOTE: Be sure to shut off power before beginning.)

1 In this first of three examples, power comes in through one switch, travels to the fixture, then to the second switch. (See page 43 for three-way switching ABCs.)

2 Locate the "common" terminal on each switch. It will be labeled or darker than the other two. Connect the incoming hot wire.

3 Attach "traveler" wires to the other terminals, then route them through the ceiling box to the other switch. Note that at the light, one white wire becomes hot.

4 Complete the hot leg of the circuit by connecting the black wire to the second switch's common terminal and to the fixture.

5 Run the neutral leg back to the power source. Either switch will operate the light.

4

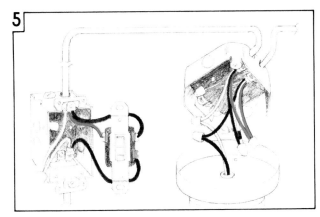

5

Installing Three-way Switches—Case No. 2

The preceding page shows how to wire three-way switches when the light is between the two switches. Here we show what to do when it's beyond both switches. We've included a dimmer in this example. (Remember: With most dimmers you can use only one per circuit.)

(NOTE: Be sure to shut off power before beginning.)

1 For this project you need to run three-conductor cable only between the switches. Power comes into the first switch and out of the second on just two wires. Connect ground wires as shown. (See opposite page for three-way switching rules.

2 As before, connect the hot leg of the power source to the first switch's common terminal.

3 Next, connect the travelers to each of the switches. (Warning: You can burn out a three-way dimmer by hooking it up incorrectly. Professional electricians often set up the circuit with ordinary three-way switches, turn on the power and check it, then replace one with a dimmer.)

4 Connect the outgoing black wire to the dimmer's common lead and to the black fixture lead.

5 Complete the circuit by connecting the white wires as shown.

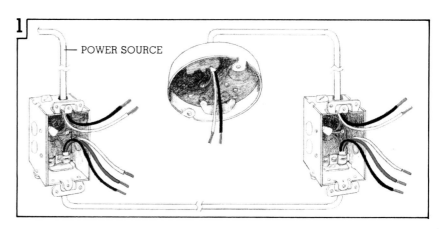

POWER SOURCE

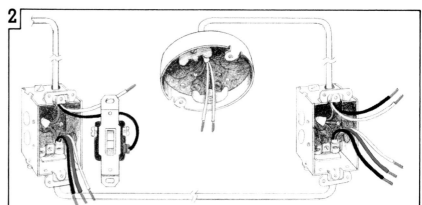

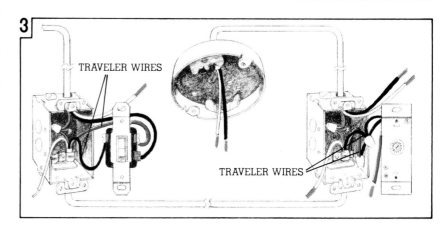

TRAVELER WIRES

TRAVELER WIRES

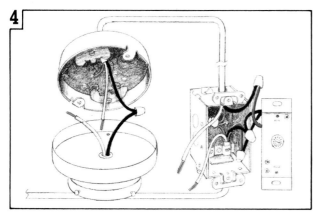

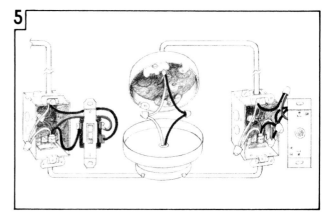

Installing Three-way Switches—Case No. 3

In this final three-way switch situation, the light fixture is ahead of both of the switches that control it.

(NOTE: Be sure to shut off power before beginning.)

1 You need three conductors only between the switches, not to and from the fixture. Connect ground wires as shown.

2 Bring the hot leg of the circuit through the ceiling outlet to the common screw of the first switch.

3 Connect the traveler wires as shown. Be sure to tape the whites.

4 Bring the hot leg back by connecting the black wire to the common terminal of the second switch and to the white wire in the first switch box. Tape it black, too.

5 Finally, connect the fixture leads as shown.

You won't get in trouble with any three-way switch installation as long as you observe these ABCs:

(A) Always attach the incoming hot (black) wire to the common terminal of one switch.

(B) Connect the traveler terminals only to each other, never to the light.

(C) Connect the common terminal of the second switch only to the black fixture lead.

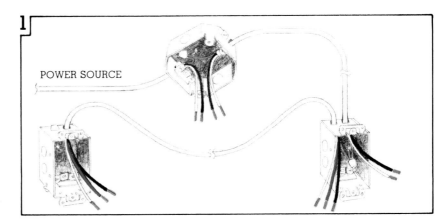

POWER SOURCE

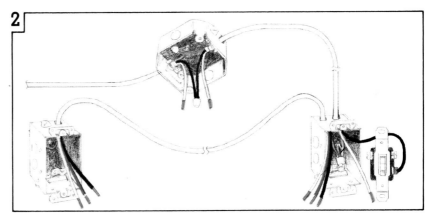

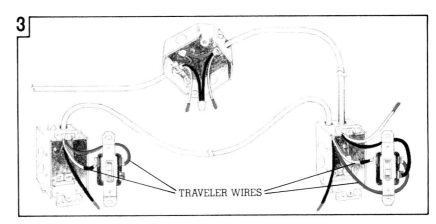

TRAVELER WIRES

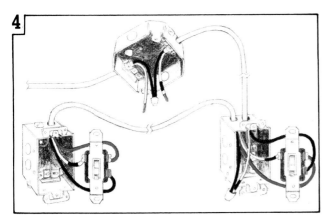

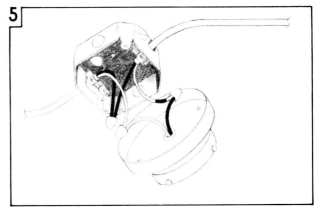

Installing Four-way Switches

And now for the final entry in our switching numbers game: the four-way switch. This one has four terminals, and you can install any number of them between a couple of three-way switches. Use four-ways when you want to control a light from three or more locations. Just remember that in four-way situations, the first and last switches must always be the three-way type.

(NOTE: Be sure to shut off power before beginning.)

1 Here, incoming power flows from switch to switch to switch to light, but it could just as easily follow one of the other routes illustrated on the preceding pages. Start by connecting the ground wires as shown. Next connect the black wire from the power source to the first switch's common terminal.

2 Now connect all the traveler wires as shown. Four-pole switches carry only travelers.

3 At the second three-way switch, link its common terminal to the fixture's black lead. This completes the circuit's hot leg.

4 Connect all the white wires back to the power source, turn on the power, and admire your switching handiwork.

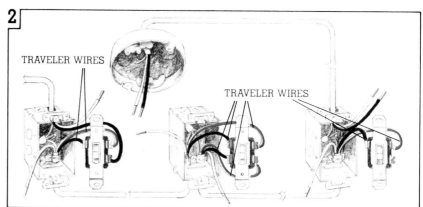

POWER SOURCE

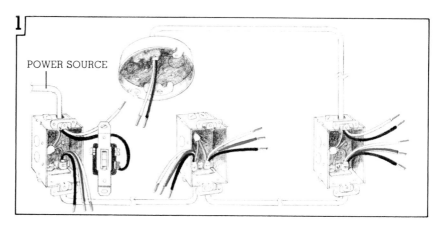

TRAVELER WIRES

TRAVELER WIRES

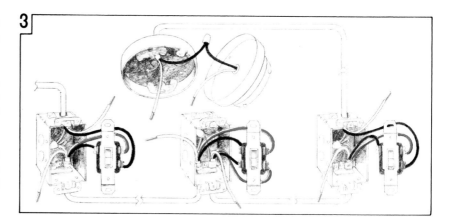

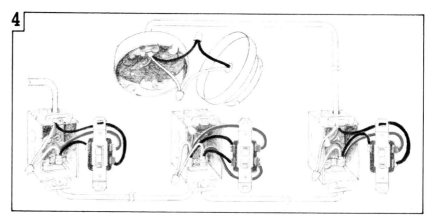

Wiring a Smoke Detector

It's true that where there's smoke there's fire, but in most cases the smoke is thickest when the fire is in its early, smoldering phase. Install an electric nose to sniff out the smoke and you gain precious minutes to do something about the fire.

As you may already know, not all smoke detectors need to be connected to house wiring. Battery-powered types mount anywhere, and give protection even during a power failure or an electrical fire. Of course you do have to replace the batteries periodically.

With a permanently wired smoke detector, you needn't worry about changing batteries. (Some units include a self-charging, battery-operated backup system that takes over if there's an electrical mishap.)

Detectors draw so little amperage that you can easily tap into an existing circuit. First, analyze your home's air currents and figure out the optimum location for a sensor. If yours is a single-level home and the bedrooms are clustered together, you probably can get by with just one unit between living and sleeping areas. Otherwise, you'll need at least two, and you might want to add more.

1 Most detectors have a mounting plate that attaches to a standard octagonal ceiling box, as shown here. Ideally, the box should be in the center of the ceiling, or on a wall about ten inches below the ceiling line.

Start by shutting off the circuit you'll be working with. Then run two-wire cable to the new outlet from a nearby junction box or receptacle, as explained on page 35. You also might be able to tap into a ceiling light fixture outlet, provided power comes to the fixture first, as in the examples on pages 38 and 40.

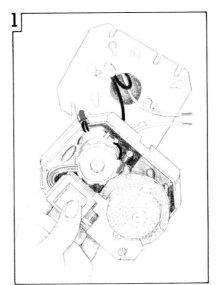

2 After making the electrical connections, feed wires back into the box and secure the unit's chassis.

The smoke detector shown, like most, ionizes air passing through, giving it a positive charge. Smoke particles interrupt the current flow and sound the alarm. Another type has a photoelectric eye that's more sensitive to heavy smoke.

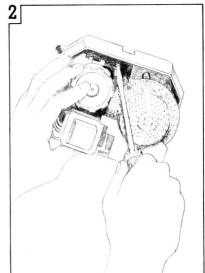

3 Now just snap on the cover, turn the power back on, and try out your new sniffer. It may have a test button. Or conduct a test by blowing cigarette smoke or holding a smoldering match near the unit. If it's working, the buzzer will sound.

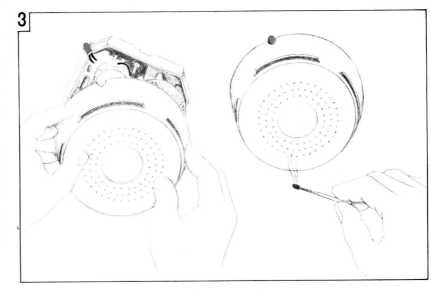

Installing a Ground Fault Circuit Interrupter

Fuses and circuit breakers protect the *wiring* in your home. Now let's meet a device that protects *you*. Called a ground fault circuit interrupter (GFCI), it senses tiny leakages of current and shuts off the power before you can say GFCI.

In most circumstances, leakage of current isn't a big problem. In properly grounded systems most of it is carried back to the service panel, and what remains would scarcely give you a tickle.

But let's say *you* happen to be well-grounded, standing on a wet lawn, for example, or touching a plumbing component. Then that tiny bit of errant current would pass through your body on its way to the earth. As little as 1/5 of an amp, about enough to light a 25-watt bulb, can be fatal.

Wired into both of the conductors in a circuit, a ground fault circuit interrupter continuously compares current levels flowing through the "hot" and "neutral" sides. These should always be equal. If the mechanism senses a difference of just 1/200 of an amp, it trips the circuit. Power is interrupted in 1/40 of a second or less, fast enough to prevent injury to any healthy child or adult.

Any ground fault is a potential hazard. Even if you're using a faulty tool or appliance with a grounding wire that's in good condition, you aren't completely safe from a serious shock.

This is why today's codes require GFCIs that monitor all receptacles in outdoor, bathroom, and garage locations —the places where you're most likely to come in contact with a

good ground. The National Electrical Code doesn't say anything about basement shops, laundries, or other heavy-equipment areas, but you might want to add a GFCI in these locations as well. A wet basement floor can provide a fatal ground.

If yours is an older home, chances are it doesn't have GFCI protection, and you should seriously consider providing it. These devices are fairly costly, but they're no more difficult to install than an ordinary receptacle or circuit breaker. The next page tells how.

(NOTE: Be sure to shut off power before beginning.)

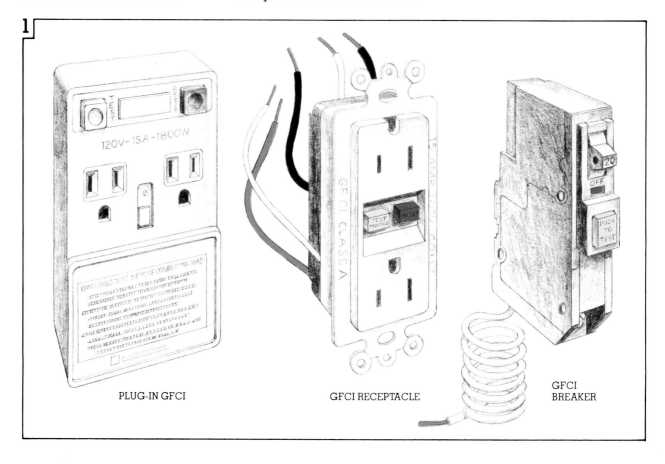

PLUG-IN GFCI   GFCI RECEPTACLE   GFCI BREAKER

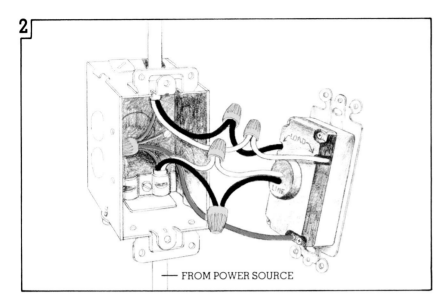

2

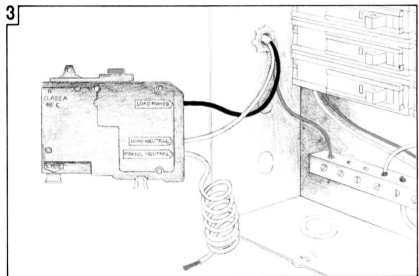

— FROM POWER SOURCE

3

4

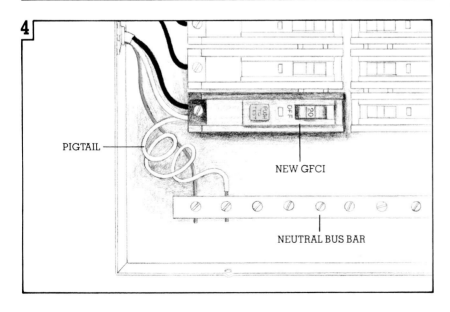

PIGTAIL —

NEW GFCI

NEUTRAL BUS BAR

1 GFCIs come in three different models. A *plug-in* type is the easiest to install. You just insert its blades into a receptacle, then plug a tool or appliance into its outlets. This one can only be used indoors, but it's portable, so you can easily move it to a work site.

GFCI *receptacles* replace ordinary receptacles and can be installed outdoors. Some protect only themselves. Others monitor the entire circuit. Test and reset buttons let you check the device.

GFCI *breakers* protect the entire circuit, too. These units also do the job of an ordinary breaker.

2 Hook up a GFCI receptacle as shown here. Incoming power goes to the *line* leads. *Load* leads send it to other receptacles on the circuit. If you're wiring one at the end of a line, cap off the load leads with solderless connectors, or buy a version that protects just one receptacle.

3 GFCI breakers clip into a service panel just as the ordinary breakers shown on page 59, but you wire them differently. To replace a conventional type with a GFCI, first shut off all power to the service panel, then snap out the old breaker.

You'll notice that only the circuit's "hot" side is connected to the breaker. The white wire goes to the panel's neutral bus bar. Disconnect *both* the hot and neutral wires and attach them to the GFCI as shown here.

4 Now ground the white *pigtail* by loosening a terminal screw on the bus bar, inserting the wire, and tightening the screw.

Finally, turn the power back on, set the breaker, and push the test button. This simulates a ground-fault condition and the breaker should trip. (Note: Follow the safety procedures on pages 58 and 59 for service panel work.)

Running Wires Underground

Have you been thinking about how nice it would be to have a post light in the front yard, power in the garage, or an outdoor receptacle for the electric mower? Go underground.

Subterranean wiring uses the same principles you've learned for inside jobs. The components differ slightly, however, and you must observe a few more precautions to protect wiring against moisture.

In planning an underground installation, first find out whether local codes permit plastic-clad UF cable, or require that you run TW wire in rigid conduit (see pages 60-61 for information on cable types). Remember that even cable must be protected with conduit where it's aboveground.

Next, decide whether you can extend power from an existing circuit, as explained on page 35, or whether a new circuit might make better sense. (For more information about new circuits, see pages 56-59.)

When you buy fittings, make sure they're the weathertight type designed for outdoor use. Exterior components look much like the ones used inside, but they're heftier and have gaskets, waterproof covers, and rubber-sealed connections.

Finally, bear in mind that because the potential for serious shock is much greater outdoors, you must protect all exterior receptacles with ground fault circuit interrupters. Pages 46 and 47 show how to install them.

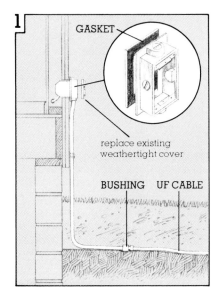

1

GASKET

replace existing weathertight cover

BUSHING   UF CABLE

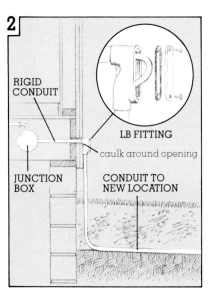

2

RIGID CONDUIT

LB FITTING

caulk around opening

JUNCTION BOX

CONDUIT TO NEW LOCATION

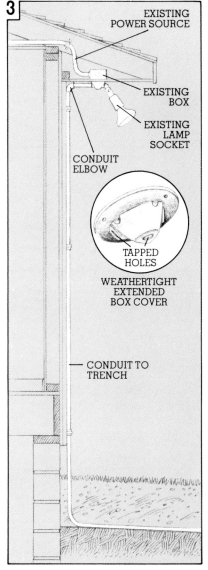

3

EXISTING POWER SOURCE

EXISTING BOX

EXISTING LAMP SOCKET

CONDUIT ELBOW

TAPPED HOLES

WEATHERTIGHT EXTENDED BOX COVER

CONDUIT TO TRENCH

1 Thinking of tapping into an exterior receptacle? Here's how you do it. Just add a weathertight *box extension* and run conduit from it. Be sure to caulk all around the extension so air and moisture can't penetrate.

If you use UF cable, it should exit from the conduit about 12 inches below grade through a special insulating bushing.

2 If no receptacle is handy, look around your basement for a junction box, or run a new circuit from the service panel.

Here, threaded rigid conduit connects the junction box to an *LB fitting*. This has a removable plate that makes it easier to pull wires through, but there's not enough space inside to make connections.

3 Although in some cases it's done, taking off from a light fixture is probably the least desirable way to go underground. You have to run unsightly conduit up to the eaves, and your new outlet will be "live" only when the light is on.

Installing a Bathroom Ventilating Fan

Flicking the switch for a bathroom exhaust fan clears the air in a hurry. A combination unit can give you instant heat and light as well. Here's how to install one.

(NOTE: Be sure to shut off power before beginning.)

1 Cut a ceiling opening between joists. Then, from the attic, nail the unit's mounting brackets to them. Make sure the fan assembly is level and flush with the surface of the ceiling.

2 If your attic has good cross-ventilation to carry off excess humidity, you can get by without venting the fan to the outside. If not, run flexible ducting to a soffit, as shown here, or to a roof-mounted outlet.

3 If your fan has a heater, as this one does, you'll probably need to run a new 15- or 20-amp circuit (see pages 56-59). Fish two-conductor cable to the switches, then run four wires from the switches to the unit's receptacles. Here's a good use for flexible metallic conduit (see pages 82-83).

For a fan-only installation you'll need just two conductors. Separate fan/light combinations require three. An existing circuit can probably handle either.

4 Now install the working parts in the housing and plug them into the appropriate receptacles. At this point you might want to turn the power on and see whether everything works properly.

Finally, fit the reflector and decorative grill. Ventilators need no more maintenance than an occasional cleaning.

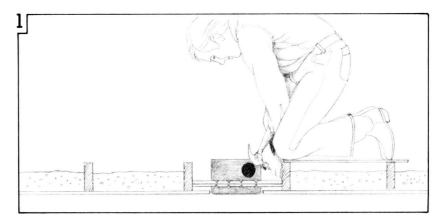

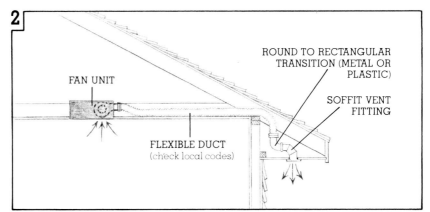

FAN UNIT

ROUND TO RECTANGULAR TRANSITION (METAL OR PLASTIC)

SOFFIT VENT FITTING

FLEXIBLE DUCT
(check local codes)

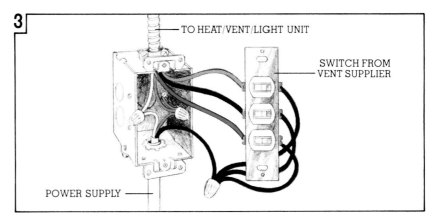

TO HEAT/VENT/LIGHT UNIT

SWITCH FROM VENT SUPPLIER

POWER SUPPLY

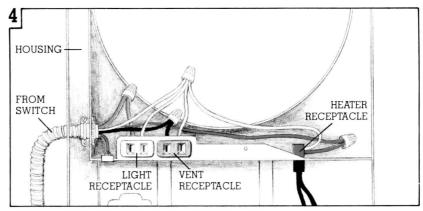

HOUSING

FROM SWITCH

HEATER RECEPTACLE

LIGHT RECEPTACLE

VENT RECEPTACLE

Installing a Power Attic Fan

On a hot summer day, the temperature in an attic can reach 150 degrees F. or more. Even after sundown, all that super-heated air continues to put a heavy strain on your home's air conditioning system. Improve attic ventilation and you can cut cooling costs by as much as 30 percent.

Turbine-type ventilators use no electricity. But because they depend on wind, you're out of luck on a still day when the heat buildup may be most intense.

Whole-house fans pull a strong, steady draft up through the attic floor. These offer an economical alternative to air conditioning in some regions and at some times of year, but you wouldn't want to run one when cooling equipment is in operation.

A power attic fan mounted in the roof or a gable-end wall lets you cool the attic and interior spaces independently of each other. Controlled by a thermostatically activated switch, it automatically kicks in to dissipate hot air. Some units also sense moisture and operate to clear away excess humidity.

To determine how big a unit you need, multiply the square footage of your attic by 0.7. Add 15 percent if your roof is dark. The numerical answer, in cubic feet per minute (CFM) tells you the size of fan you should shop for.

(NOTE: Be sure to shut off power to the circuit before beginning your installation.)

1 Locate your fan as close to the ridge as possible, and on a slope that's not visible from the street. Then get up in the attic, pick out a pair of rafters, and measure to a point midway between them. Drill up through the roof at this point.

2 On the roof, use the hole as a center point and trace an outline of the square the unit and its flashing will occupy. Check this by setting the fan in place. The flashing should tuck neatly under shingles on the up-roof edge. To minimize the number of shingles you have to trim, you may want to adjust the unit up or down slightly and find a new center point.

3 A template that comes with the fan helps you mark the circle you must cut out. After making all your measurements, double-check them, then slice through the shingles with shears or a utility knife.

In a gable-end wall, follow basically the same procedure. When you cut — you'll need a circular saw here — go through only the siding, not the sheathing underneath. If the wall has a window or louvers, you might want to adapt the opening to accommodate the ventilator.

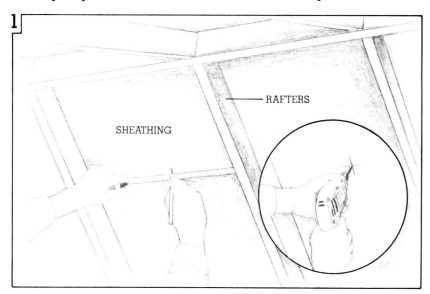

RAFTERS

SHEATHING

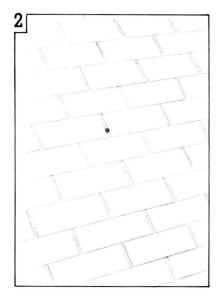

4 A second template will guide you in marking a smaller circle inside the cut-away shingles. Use a saber or keyhole saw to get through the roofing paper and sheathing. Next, remove any nails in shingles that lie in the top two-thirds of the square you've drawn. Now you're ready to mount the unit.

5 Turn the fan over and coat the underside of its flashing with roofing cement. Then slide it under shingles above and on either side; the flashing remains atop the shingles below.

Now, referring to the square as a guide, center the ventilator over the hole. Most units have a mark that indicates which edge goes on the up-roof side.

Carefully pry up each shingle and nail the flashing to the sheathing with galvanized roofing nails. With roofing cement, seal all joints, holes, and nails, and stick down the shingles you've lifted. Your attic fan is now weathertight. All that remains is a simple electrical hookup, probably from an existing circuit.

6 Back in the attic, screw the thermostat switch to a stud or rafter that's above the ventilator and out of the air stream it will create. Remove the box cover plate.

7 Run cable from a junction box to the thermostat and to a switch below that you can use to override the automatic control if you wish.

8 Turn on the power and test your installation. A screw in the thermostat box lets you adjust the temperature at which the fan will come on. It will shut off when the attic is 10 degrees cooler.

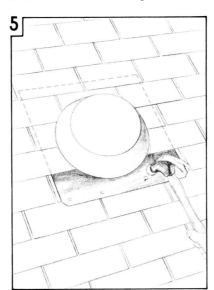

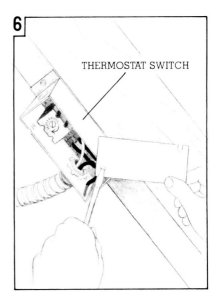

THERMOSTAT SWITCH

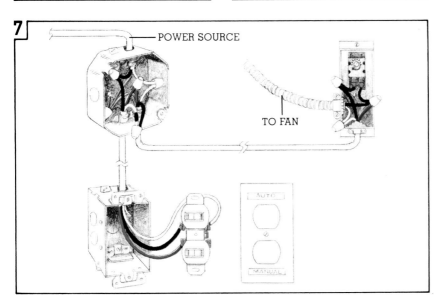

POWER SOURCE

TO FAN

AUTO

MANUAL

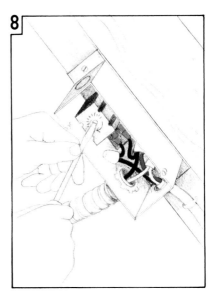

Installing an Intercom

How are communications at your house? If you find yourselves shouting from room to room, missing callers because you're too tied up to answer the door, or straining to hear a baby's cry, your family could benefit from the convenience of an intercom network.

Intercoms come with a variety of options—music systems, buzzers that unlatch doors, speaker units for almost any situation—and all operate on low-voltage current.

The drawings here and on the opposite page show how to fish low-voltage cable through finished walls and make all electrical hook-ups. Since low-voltage wires present no danger of shock or fire, you can also use the cable for exposed runs.

1 Situate the master unit at a central point, such as the kitchen or family room. It should be placed on an interior wall away from cabinets, fluorescent lights, and plumbing. Most housings fit between studs spaced 16 inches on center. Find a pair and cut open the wall as shown here.

Installers prefer a height of about 52 inches above the floor, handy to a receptacle or junction box they can tap for power. Plumb the housing and screw it to the studs on either side.

2 Locate substations near doorways and next to light switches if you can. Cut an opening for each mounting ring and fasten it to the surface of the wall.

3 Plan to fish wires through the attic or basement, whichever is easier. Bore holes from the unfinished space through plates above or below the master unit and substations. You may need several holes for the master unit.

4 Now fish cables up or down to the units. In an attic, protect the wires by running them along rafters, not joists. Take care when you're stapling that you don't pierce the cable.

Some installations require four-conductor cable, others require six-conductor cable. Check the manufacturer's instructions and use only the type specified. In situations where you can't fish, either can be easily concealed behind baseboards and other trim. Even in exposed runs, they aren't any more obtrusive than indoor telephone cable.

If your intercom has a radio, you'll also have to fish an antenna lead up to the attic. This one can't go to the basement or dangle inside a wall.

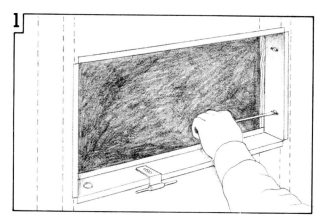

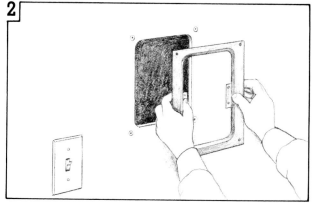

TOP PLATE ABOVE INTERCOM UNIT

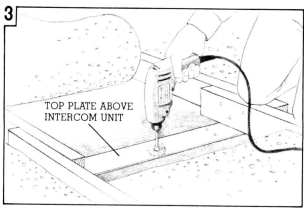

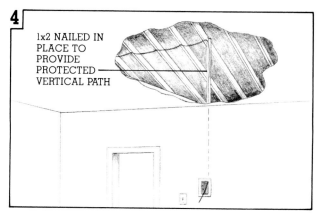

1x2 NAILED IN PLACE TO PROVIDE PROTECTED VERTICAL PATH

5,6 Now it's time to connect all those wires. Work methodically, tying in one substation at a time, as shown here.

Begin by knotting each cable where it enters the housing and stripping away its outer insulation. Color-coding helps you keep the conductors straight, but make sure each is attached securely to its screw terminals. Having to troubleshoot a wrong or loose connection after everything is hooked up is a nuisance.

7 Outdoor substations are simply speakers and need only two conductors, usually colored black and white. Attach them to the proper terminals.

8 If you have a doorbell, you might be able to use its wire for the outdoor substation. Or tie the cable to the bell wire and pull it to draw the cable through your exterior wall.

An electrically controlled latch requires a second set of conductors to the door's strike plate. A special mechanism here lets you unlock the door with a button at the master station or a substation. The door automatically locks itself again as it closes.

(NOTE: Be sure to shut off power before doing step 9.)

9 Some master units have built-in transformers. With these you shut off the circuit you're tapping, bring in 120-volt cable, and connect it within the housing.

If yours doesn't have an integral transformer, wire a separate one as illustrated. Fasten it to a junction or receptacle box and run low-voltage cable to the master station.

10 Finally, make power connections to the master unit, turn the power on, and test your installation. If you have problems, consult the manufacturer's instructions for troubleshooting procedures.

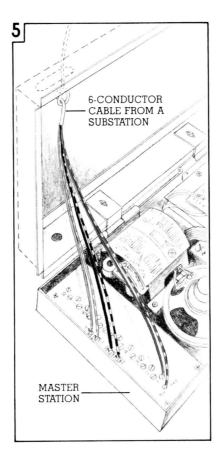

6-CONDUCTOR CABLE FROM A SUBSTATION

MASTER STATION

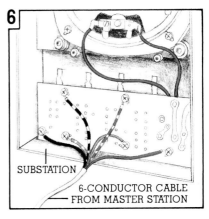

SUBSTATION

6-CONDUCTOR CABLE FROM MASTER STATION

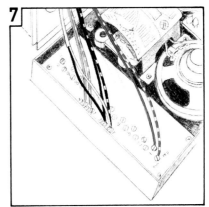

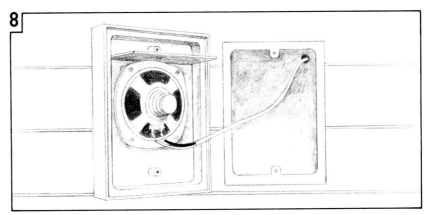

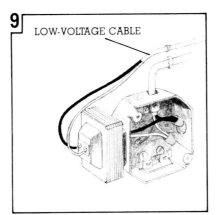

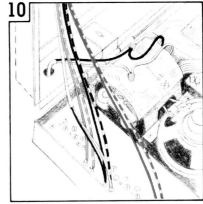

LOW-VOLTAGE CABLE

Adding Circuits to Your System

And now for the most advanced project in this book: tapping new circuits into your home's service panel. Whether you'll want to try this one yourself or hire an electrician to do all or part of the job depends as much on the situation at your service entrance as it does on your confidence and electrical expertise.

This page and the next tell how to assess that situation. The two that follow show how to make the final connections.

Your first consideration: Are the wires to your meter bringing in enough capacity to handle the extra load? Look for an amperage rating on your home's main fuse, circuit breaker, or disconnect switch. Older 60-amp service probably can't be stretched. Upgrading this is a job for your utility or a professional.

Newer 100-amp service may have enough reserve to handle an additional circuit or two, while 150- or 200-amp service usually can take care of all but a major addition to your home's electrical load. To find out exactly how much more you can add, go through this step-by-step analysis.

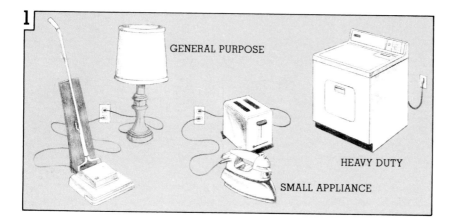

GENERAL PURPOSE

HEAVY DUTY

SMALL APPLIANCE

Circuit Need Selector

Location	Circuits
Living, dining, bedrooms, hallways, finished basements	A 15-amp general-purpose circuit for each 500 square feet. For a room air conditioner, install a small-appliance circuit.
Kitchen	At least two 20-amp small-appliance circuits and a 15-amp lighting circuit. An electric range needs a 120/240-volt circuit.
Bathroom	A 15-amp general-purpose circuit with GFCI protection.
Garage	A 15- or 20-amp general-purpose circuit with GFCI protection.
Laundry	A 20-amp small-appliance circuit for the washer and a gas dryer; you'll need a 120/240-volt circuit for an electric dryer.
Workshop	A 20-amp GFCI circuit; for larger shops run two, or a separate lighting circuit.
Outdoors	One 20-amp GFCI circuit.

1 Begin by breaking down your home's circuits into the three categories illustrated here: general-purpose, small-appliance, and heavy-duty.

General-purpose circuits may have about 10 light or receptacle outlets. They usually carry 15 amps, sometimes 20. Ideally you should have one general-purpose circuit for every 500 square feet of living space. Some local codes stipulate that lighting and receptacles be on separate general-purpose circuits.

Small-appliance circuits often supply 20-amp current to just two or three receptacles. Codes call for two small-appliance circuits in the kitchen, and you'll probably need others for the laundry, a shop, or a window air conditioner.

Heavy-duty circuits feed just one customer, such as a furnace, an electric dryer, or a range. You may need 240- or 120/240-volt wiring for some of these appliances.

Which of these circuits do you need, and where? The chart at left, based on requirements in the National Electrical Code, tells what you should have. Note that some circuits now must be equipped with ground fault circuit interrupters (GFCIs), as explained on pages 46 and 47.

Add up the amperage ratings for all of your home's circuits and you may be alarmed to discover that their total exceeds the service rating on the main fuse or circuit breaker. Does this mean your electrical system is already overtaxed?

Probably not. Remember that few if any of the circuits ever work at full amperage capacity, and it is unlikely that you'll ever operate *all* of your home's electrical equipment at the same time.

This is why codes allow electricians to *de-rate* total household demand when they're computing how much service a home needs. The table at right goes through a typical de-rating calculation for a 2,000-square-foot house with 100-amp service, five small-appliance circuits, and two heavy-duty circuits.

In assigning wattage values, don't count the number of general-purpose circuits at all. Just use three watts per square foot of house area as a rule of thumb. Small-appliance circuits count at 1,500 watts each. Only with heavy-duty circuits do you use full wattage ratings, and if two items, such as an electric furnace and air conditioning, never run at the same time, ignore the one that draws less.

Rate only the first 10,000 watts of the total at full value, then calculate 40 percent of the remainder. The answer, 64.2 amps in this example, indicates that the system easily could accommodate several more circuits.

2 Now let's see if your panel has room for any more circuits. If it has circuit breakers like the one below, you might find a blank space or two. If not, double up by replacing an existing breaker with a pair of *skinnies* or a *tandem* device.

In a fuse box you might be lucky enough to find an unused *terminal* and *socket* that could be put to work. More likely, you'll have to add a sub-panel, as shown on page 59.

How to De-rate Service Capacity	
Formula	**Example**
Add	
General-purpose circuits (square footage x 3 watts)	6,000 W
Small-appliance circuits (number x 1,500 watts)	7,500 W
Heavy-duty circuits (total of appliance name-plate ratings in watts)	10,000 W
Total:	23,500 W
Compute	
The first 10,000 watts at 100%	10,000 W
The remaining 13,500 watts at 40%	5,400 W
De-rated total:	15,400 W
Divide	
The total de-rated wattage by voltage	$\dfrac{15{,}400 \text{ W}}{240 \text{ V}}$
De-rated amperage:	64.2 A

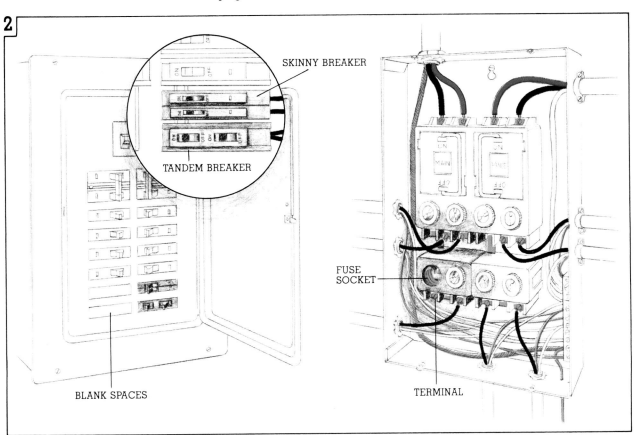

2

SKINNY BREAKER

TANDEM BREAKER

BLANK SPACES

FUSE SOCKET

TERMINAL

Adding Circuits
(continued)

3 To install a new circuit, you work backwards, first mounting boxes for the new outlets, then interconnecting them with cable and bringing it back to the service panel in what the pros call a "home run." Pages 35-44 and 60-89 show what happens in the outfield and at the bases. Here's the scene at home plate.

(**NOTE: Before you even remove the cover from a service panel, make absolutely certain that no power is flowing through it or to it.**) Where is your home's main disconnect switch, fuse, or circuit breaker? It might be outside near the meter (as shown here), inside near the service panel, or in the panel.

If you have a remote disconnect, you can just flip off the switch or pull the fuse, open up the box, and test it as explained below. If, however, your main disconnect is part of the service panel, you should seek advice from an electrician, an electrical inspector, or your utility company. Turning off an integral disconnect kills power to the individual circuits, but live current will still be coming into the box. To stop the flow entirely, you may have to have the meter pulled, as illustrated, or hire an electrician to make the final connections.

4 Now let's make sure the power is off. Standing on a board, rubber mat, or other insulator, and being careful not to touch electrical or plumbing fittings, remove the cover plate.

Then, identify the terminal screws for the main power cables. They'll be the biggest ones in the box and probably will be located at the top. Gently touch them with the probes of a voltmeter, as illustrated here. If you get a reading, the box is still live. Here, both lines are still carrying power to the breaker panel; at the fuse panel, both are dead.

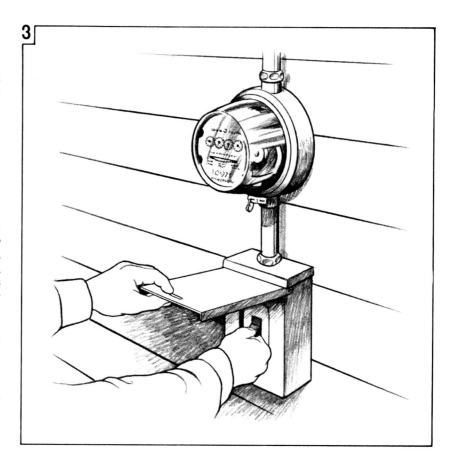

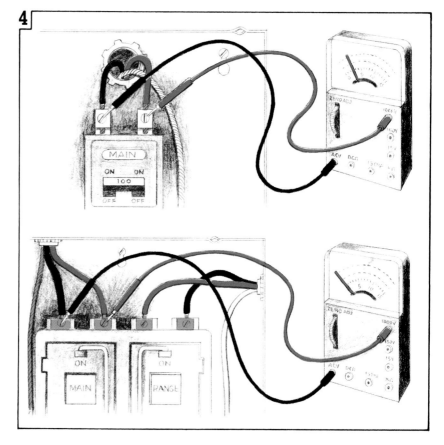

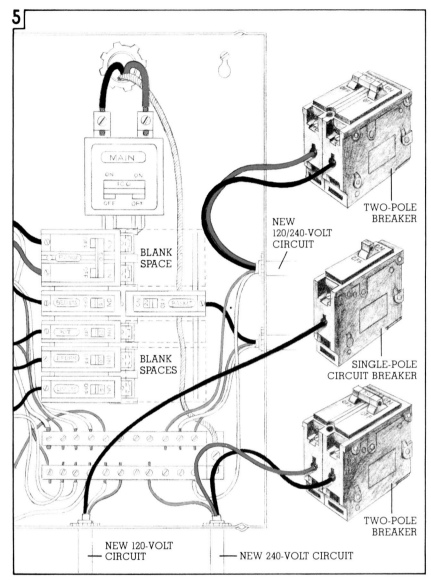

5
TWO-POLE BREAKER

NEW 120/240-VOLT CIRCUIT

SINGLE-POLE CIRCUIT BREAKER

TWO-POLE BREAKER

MAIN

BLANK SPACE

BLANK SPACES

NEW 120-VOLT CIRCUIT

NEW 240-VOLT CIRCUIT

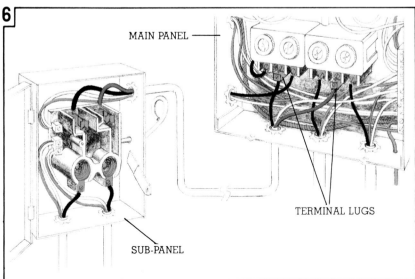

6
MAIN PANEL

TERMINAL LUGS

SUB-PANEL

5 When you're satisfied the power is off, you can make relatively quick work of the final connections. If you're hooking in several circuits, connect them one at a time so you can keep the wires straight.

Begin by punching the center from a convenient knockout. Then pry out one or more of the knockout's concentric rings to make a hole the size of the cable connector you'll be using.

Now strip back the cable covering, allowing for enough wire to reach the neutral bus bar as well as the spot you have in mind for the new breaker. Connect the cable to the box.

Inside, run the white and circuit ground wires to the neutral bus bar. For a 120-volt circuit, attach the red or black wire to the terminal of a *single-pole* breaker (see sketch). Clip the breaker onto one of the panel's hot bus bars and the job is done.

For a 240-volt circuit, which is really just two 120s, you need a *two-pole* breaker. It's twice as wide as a single-pole type. In this case, both of the circuit wires are hot and attach to the breaker as shown. Only the ground wire goes to the neutral bus bar.

Combination 120/240-volt circuits use the same two-pole breaker; connect the third, white wire to the neutral bus bar. GFCI breakers install somewhat differently, as explained on page 47.

6 A sub-panel gets you out of a crowded fuse box, and if you're adding several circuits far from the main panel, you might want to connect them to a remote sub, then make just one "home run" back to the main panel.

Besides the new panel itself and the fuses or breakers it will hold, you also need cable with three wires sized to handle the amperage the sub will draw (see page 61), plus the *terminal lugs* shown.

ELECTRICAL BASICS AND PROCEDURES

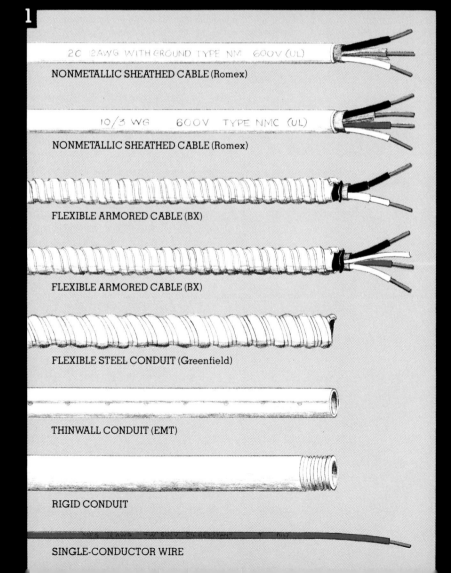

1

2C 12AWG WITH GROUND TYPE NM 600V (UL)

NONMETALLIC SHEATHED CABLE (Romex)

10/3 WG   600V   TYPE NMC (UL)

NONMETALLIC SHEATHED CABLE (Romex)

FLEXIBLE ARMORED CABLE (BX)

FLEXIBLE ARMORED CABLE (BX)

FLEXIBLE STEEL CONDUIT (Greenfield)

THINWALL CONDUIT (EMT)

RIGID CONDUIT

SINGLE-CONDUCTOR WIRE

Once you've decided where you want a new outlet and how you're going to tap into an existing circuit or the service panel itself, you're faced with the task of actually getting the project done. This section gives the nuts-and-bolts know-how you need to do just that.

Electrical projects amount to little more than simple assembly work. You just fasten together a series of standardized components in equally standardized ways.

The pages that follow present the cables, boxes, fittings, and other devices you'll be using, and tell how the National Electrical Code says they are to be assembled.

Don't, however, rely solely on the national code. It is simply a series of widely accepted recommendations about how work should be done. Local codes take precedence, and these vary in some details from community to community.

Not all codes let homeowners do their own wiring. Some require that a licensed person make the final connections. Others allow you to do the work after passing a test that shows you're familiar with the basics explained here.

Selecting Cable

1 Most local codes let you use *nonmetallic sheathed cable* inside walls, floors, and other places where it can't be damaged. Information on the plastic covering tells what's inside.

Our topmost example has two No. 12 (American Wire Gauge) conductors, plus a ground. *Type NM*, also called Romex, a trade name, goes in dry locations only.

The second cable shown has three No. 10 conductors, plus a ground. *Type NMC* runs in dry or damp locations. A third type of nonmetallic cable, not shown, is *UF* (underground feeder.) It's allowed in wet places, too, which means you can bury it.

Flexible armored cable, also known as *BX*, another trade name, has a spiral-wrapped steel cover. It can be used in dry or damp places and for short exposed runs.

Flexible steel conduit, or Greenfield, looks like armored cable without the wires. You cut it to length, thread wires through it, then install the completed pieces. With *thinwall* and *rigid conduit*, you fish wires after conduit has been run. Most codes require you to use these for exposed runs.

2 In some localities, you can save money by running *aluminum* or *copper-clad aluminum* wiring. Be warned, however, that aluminum expands and contracts considerably more than copper, causing loosening at terminals. So use only devices marked *CO/ALR* or *CU/AL* with aluminum.

3 The size of wire you must install depends on the amperage it will carry. Wire sizes are inversely proportional to their numbers. Drawings at right show the sizes you'll likely work with and some typical uses. Use one size larger with aluminum.

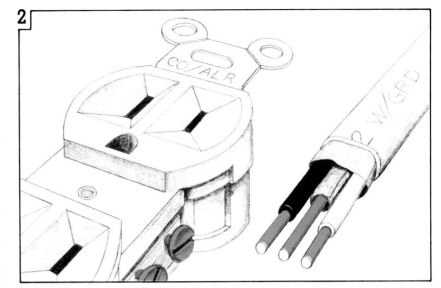

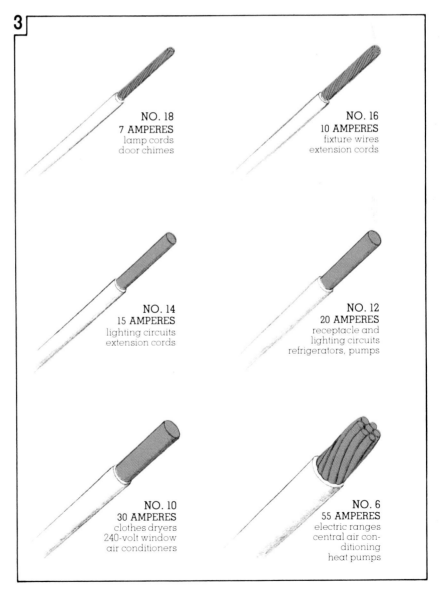

NO. 18
7 AMPERES
lamp cords
door chimes

NO. 16
10 AMPERES
fixture wires
extension cords

NO. 14
15 AMPERES
lighting circuits
extension cords

NO. 12
20 AMPERES
receptacle and
lighting circuits
refrigerators, pumps

NO. 10
30 AMPERES
clothes dryers
240-volt window
air conditioners

NO. 6
55 AMPERES
electric ranges
central air conditioning
heat pumps

Electrical Boxes, Accessories, and How to Install Them

What's Available in Boxes

Shop for electrical supplies and you won't find many cables that differ from the ones illustrated on page 60. The drawings on the opposite page, on the other hand, show only a few of the hundreds of different boxes and accessories you can choose from.

When you think about the jobs a box might be called on to do, you begin to understand why so many versions are available. Primarily, of course, any box has only one function: to house electrical connections. But those connections might be to a switch, a receptacle, the leads from a light fixture, or other sets of wires.

Codes govern how many connections you're allowed to make within a box, depending on its size. If you must make more, you have to use a bigger box. (See the chart below.)

Another major reason for the variety of boxes is convenience. If, for instance, you'll be pulling cables through a finished wall, use an *old-work* box that can be mounted from outside the wall.

To keep confusion at a minimum, let's break them down into a few broad categories.

1 *Switch/receptacle* boxes hold either device and serve as the workhorses in any electrical installation. Several of the metal ones shown here can be ganged into double, triple, or larger multiples by removing one side.

Take note of the round-cornered version shown. Called a *utility* or *handy box*, it is used with conduit in exposed locations and cannot be ganged.

Nonmetallic switch/receptacle boxes are made of plastic and are prohibited by some codes. They can't be ganged. *Weatherproof boxes* go outdoors. More about these on pages 48-50.

2 *Fixture/junction* boxes may support lighting fixtures, or split circuits off into separate branches. These are round, octagonal, or square.

3 *Accessories*, as you can see from the sample array here, are myriad. You can easily figure out what most of these do, or check the pages to come.

Determining Which Box to Use

Overcrowd a box and you risk damaging solderless connectors, piercing insulation, or cracking a switch or receptacle, any of which could cause a short. That is why codes spell out how many wires you may install in boxes.

The table at left displays the NEC requirements. As you count, bear in mind these points:
1) Don't count fixture leads joined to wires in the box.
2) Count as one a wire that enters and leaves without a splice.
3) Count as one any number of cable clamps, hickeys, or studs.
4) Don't count external connectors, but if the box has internal clamps, do count them as one.
5) Count each receptacle or switch as one.
6) Grounding wires running into a box also count as one.
7) Don't count grounding wires that begin and end in the box.
(continued)

Choosing the Correct Box Size

Type of box	Size in inches (Height x width x depth)	Maximum number of wires			
		No. 14	No. 12	No. 10	No. 8
Switch/ Receptacle	3x2x1½	3	3	3	2
	3x2x2	5	4	4	3
	3x2x2¼	5	4	4	3
	3x2x2½	6	5	5	4
	3x2x2¾	7	6	5	4
	3x2x3½	9	8	7	6
Utility	4x2⅛x1½	5	4	4	3
	4x2⅛x1⅞	6	5	5	4
	4x2⅛x2⅛	7	6	5	4
Fixture/ junction	4x1¼ round or	6	5	5	4
	4x1½ octagonal	7	6	6	5
	4x2⅛	10	9	8	7
	4x1¼ square	9	8	7	6
	4x1½ square	10	9	8	7
	4x2⅛ square	15	13	12	10
	4¹¹/₁₆x1½ square	14	13	11	9
	4¹¹/₁₆x2⅛ square	21	18	16	14

1 SWITCH/RECEPTACLE BOXES

METALLIC

NONMETALLIC

WEATHER-PROOF

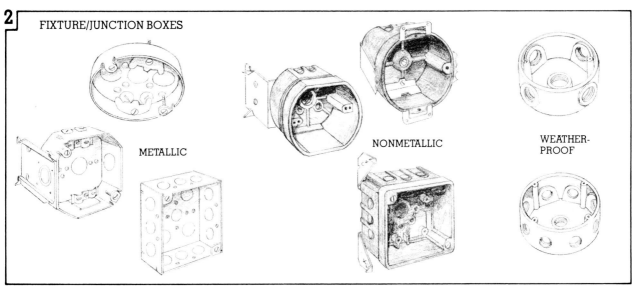

2 FIXTURE/JUNCTION BOXES

METALLIC

NONMETALLIC

WEATHER-PROOF

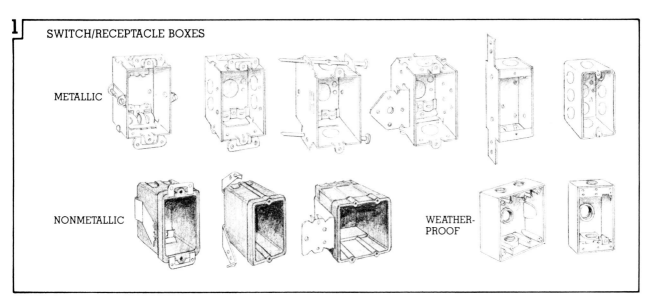

3 ACCESSORIES

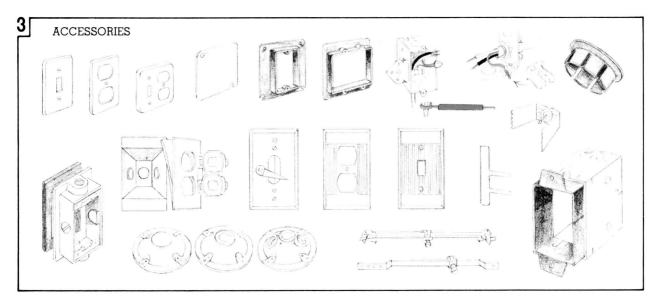

Electrical Boxes
(continued)

Installing Boxes in Unfinished Space

The order you decide to follow in doing the procedures shown here and on the next 20 pages depends largely on the state of the walls and ceilings in which you'll be adding outlets.

When electricians wire a new house, they typically fasten boxes to the framing, as illustrated here, run cable from box to box, then make home runs to the service panel. You'd do the same in a new addition, basement remodeling, or other unfinished construction project. After you've "roughed in" your wiring, you put up wall and ceiling materials, then install devices in the boxes.

But when you have to fish wires through finished walls and ceilings, matters become more complicated. Then you usually begin by making an opening where you want the new outlet and bringing cable to the opening. Only after you've fished the cable and connected it to the box do you install the box.

If you're wiring unfinished space, count yourself lucky. With the framing out in the open, you can install a box in about the time it takes to drive a couple of nails. Here's how:

1 With switch and receptacle outlets, the critical thing to keep in mind is that the box's edge must end up flush with the surface of the wall you'll be installing. New-work boxes let you compensate for the thickness of paneling or drywall in one of several ways.

Many *nail-up* and *straight-bracket* boxes have a series of *gauging notches* on their sides. You align the appropriate notch with the outer edge of the stud and nail or screw it in place. Note that the straight-bracket box shown here is only 1½ inches deep, so it can be attached to a 2x2 or 2x3 furring strip.

Some *L-bracket* boxes adjust to suit the thickness of your wall material. Others accommodate only one thickness. *Utility (handy) boxes* mount on the surface, so you needn't worry about aligning them. Run conduit first (see pages 78-81), and attach to masonry as shown on the opposite page.

Locate switches 48 to 50 inches above the floor and receptacles 12 to 16 inches above it. Check what your local code says about spacing between receptacles. The National Electrical Code requires that they be placed so that no point along any wall is more than six feet from an outlet. This means a receptacle for every 12 running feet of wall.

1

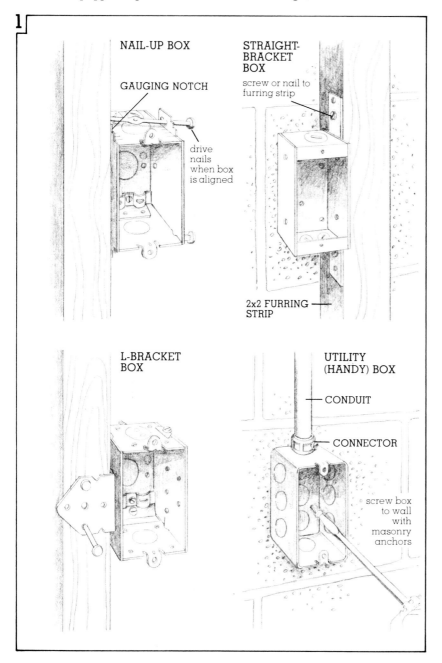

NAIL-UP BOX

GAUGING NOTCH

drive nails when box is aligned

STRAIGHT-BRACKET BOX

screw or nail to furring strip

2x2 FURRING STRIP

L-BRACKET BOX

UTILITY (HANDY) BOX

CONDUIT

CONNECTOR

screw box to wall with masonry anchors

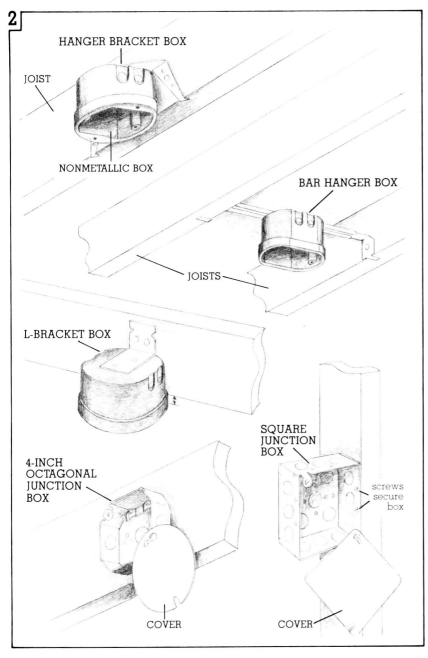

2 HANGER BRACKET BOX

JOIST

NONMETALLIC BOX

BAR HANGER BOX

JOISTS

L-BRACKET BOX

SQUARE JUNCTION BOX

screws secure box

4-INCH OCTAGONAL JUNCTION BOX

COVER

COVER

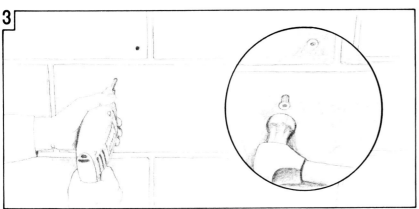

3

2 Fixture and junction boxes are equally easy to install at the framing-in stage. As with switch and receptacle outlets, all codes demand that fixture boxes be flush with the finished ceiling or wall surface. Junction boxes, on the other hand, are often covered. They go inside wall or ceiling cavities.

Is there a joist at the spot where you want a fixture? Boxes with a *hanger bracket* fasten to the joist as shown. Be sure to allow for the thickness of the ceiling material you'll be using. You can buy metal boxes with brackets similar to the one on this plastic version.

A *bar hanger* lets you locate a box between joists. This one adjusts to suit different joist spacings, and you can slide the box along the bar to situate your new outlet where you want it.

An *L-bracket* offers yet another way to go. Fixture boxes don't have gauging notches, so in an installation like this one you have to measure to compensate for the ceiling material's thickness.

Some junction boxes come with brackets. Others just nail or screw to a joist, stud, or rafter. Regardless of the type of box you're installing, always secure it with two fasteners. And if the box will be supporting a light fixture or other heavy object, make sure it's anchored well enough to carry the load.

3 Mounting a box on a masonry wall calls for a bit more work. Hold the box in position and mark the location of its two holes. Next, drill into the wall with a masonry bit and insert a couple of lead or plastic screw anchors. Finally, screw the box to the anchors. *(continued)*

Electrical Boxes
(continued)

Installing Boxes in Finished Space

The preceding two pages show the simple procedures involved in installing boxes in unfinished space. But most of us have to deal with the reality of finished walls and ceilings, a more challenging prospect.

How do you get at the framing and attach boxes to it without making big holes? In most instances, you don't want to make contact with the framing. Instead, simply make a hole the size of the box, then secure it to the surface. Special old-work boxes and accessories make the job easier than you might envision.

Before you begin, though, consider how you're going to get cable to the new location. Pages 73-76 show the routes you should explore.

1 Begin by making a small test hole, inserting a bent piece of wire into the wall, and rotating it as shown. If the wire hits something, move a few inches and try again. In some walls fireblocking stretches horizontally between the studs at a height of about 48 inches, just where you'd like to locate a switch. If this is the case, you'll have to install the box higher or notch it into the blocking.

2 Some old-work boxes come with a template to hold against the surface and trace around. Otherwise, use the box itself. Just make sure the box or template is level before you mark the outline of the opening.

3 Now carefully cut around the outline. If the surface is drywall, you can do this with a utility knife. In paneling or plaster, use a keyhole saw. With plaster you'll encounter wood or metal lath that also must be cut. Mask around the outline with tape to prevent crumbling.

4 The wall opening serves as a hole in the ice for your cable-fishing expedition. Once you've pulled in the cable, attach it to the box (see page 77) and fit the box in place.

To secure it, you have several options. If the opening has paneling or sound lath around it, you probably can fasten the box's ears to the wood with screws.

Box support straps tie metal switch/receptacle boxes securely to drywall and plaster/metal lath walls. *Side-clamp* boxes grip the surface from behind when you tighten the screws. Some plastic boxes have a *spring clip*. Push the box into the opening up to its ears and the clip pops open inside the wall cavity.

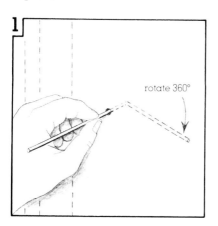

rotate 360°

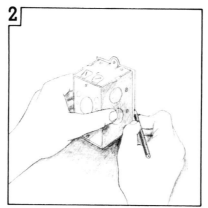

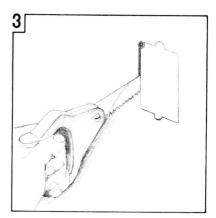

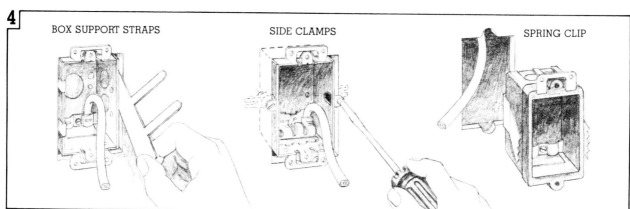

BOX SUPPORT STRAPS   SIDE CLAMPS   SPRING CLIP

5 A ceiling box has to support a fixture as well as itself, so in most cases you should fasten directly to framing. If you're fortunate enough to have attic space above, here are three ways to attach to joists.

First, mark the box's location on the ceiling and drive nails up through it to orient you upstairs. If there's a joist adjacent to the opening, nail an L-bracket box to it. If not, use a *bar hanger* or 2x4 support.

6 No access from above? Then you'll have to try one of these alternatives for your installation.

If the fixture is lightweight, such as a smoke detector, you might be able to get by with a *spring-clip* box that works just like the switch/receptacle version shown opposite. *Ceiling pan* boxes are only ½ inch deep and recess into the surface.

As a last resort, you may have to open up the ceiling and install a *bar hanger*. With drywall, cut out

a rectangle. With plaster, chip a path and install an *offset hanger*.

7 After you've checked the electrical installation, patch up the ceiling opening as shown here. With drywall, you might be able to use the same piece you cut out. Make an opening for the box, nail the panel to the joists, and tape the seam with joint compound. Fill plaster with patching compound.

5

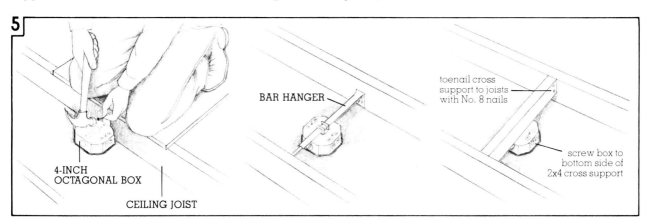

4-INCH OCTAGONAL BOX

CEILING JOIST

BAR HANGER

toenail cross support to joists with No. 8 nails

screw box to bottom side of 2x4 cross support

6

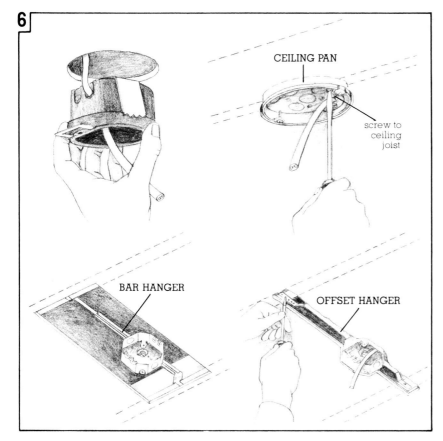

CEILING PAN

screw to ceiling joist

BAR HANGER

OFFSET HANGER

7

How to Work With Nonmetallic Sheathed Cable

The Paths Cable Can Travel

When is a straight line not necessarily the shortest distance between two points? Answer: When the line is an electrical cable zigzagging from one outlet to another in a circuit.

Electricity basically *wants* to zip along conductors when there's a call for its energy. Sharp turns and longer trips don't bother it. This means you can snake cable up and down walls, along or across joists, and around obstructions without worrying that you'll impede the flow.

Cable is priced by the foot, however, and extra feet can add up fast. So to be economical, you'll want to keep your runs as short as possible.

In new work, that's not too difficult. Examine the cutaway of a typical new-home installation at right, and you can see that most of the cables proceed directly to their destination.

Locating the dryer near the service panel, for instance, let the electrician minimize the amount of heavier, more costly cable needed for its 240-volt circuit, and he didn't have to bend and install much conduit to carry the wire.

Here two general-purpose circuits pop up through the *sole plate* and take off around perimeter walls to receptacle outlets. Others follow or span floor joists. Note that a junction box such as the one shown in the foreground lets you save a considerable amount of money and effort by bringing a single three-conductor cable from the service panel to the box, then splitting off with two-wire cable where the circuits diverge.

In planning an electrical layout, especially if you'll be running more than one circuit, it helps to draw a floor plan of your home to scale, then mark in the routes cable will traverse. To estimate how much you'll need, measure the distances involved, add 10 percent for bends and unexpected detours, and another six to eight inches for each time cable will have to enter or leave a box.

If you're thinking about fishing through finished walls and ceilings, you have some detective work ahead of you, and you'll probably use more cable than you would if the framing were exposed.

Your first task is to determine exactly what's in the space you want to run cable through. Let's say, for instance, that you want to install a new appliance receptacle for an air conditioner in an exterior wall, directly above the service panel. If there's no insulation in that wall, it's easy to drop straight down to the panel. If the wall is insulated, though, the task becomes more difficult, because you'll have to work your way through the obstruction. (You'll find some tricks for wiring finished spaces on pages 73-76.)

Finally, if your project will include switches, especially three-ways, brush up on switching basics (pages 37-44), and indicate on your diagram where you'll need to run three-wire cable. *(continued)*

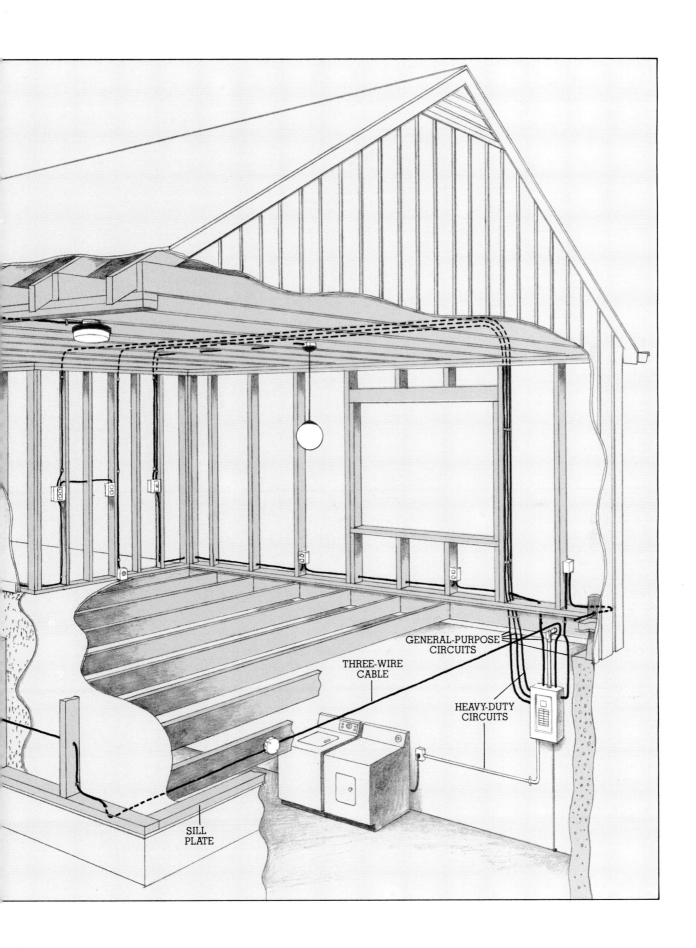

GENERAL-PURPOSE
CIRCUITS

THREE-WIRE
CABLE

HEAVY-DUTY
CIRCUITS

SILL
PLATE

Running Cable in Unfinished Space

Horizontally Through Wall Cavities

Here you have two choices: Notch the studs and run cable through the notches, as illustrated on this page, or bore holes through the studs, as shown on the opposite page.

With interior partitions you can choose either option, but you might find boring easier. On outside walls notching the studs is best. Running cables through studs would compress the insulation, reducing its effectiveness.

1 With a chisel, make a notch in each stud that's slightly larger than the cable you'll be using. At corners, two notches get you around the bend.

2 Protect cable at each notch with a 1/16-inch-thick *steel plate.* Its purpose is to ward off any nails or screws that might be driven accidentally through the wall surface, penetrate the cable, and cause a short in the circuit.

Plates like the ones shown in this sketch, which are available from electrical suppliers, have spurs top and bottom. Just drive them into the wood with a couple of hammer blows.

3 Don't try to make sharp bends with cable. Crimping it could break a conductor or damage the insulation. Instead, arc gradually around corners and up to boxes.

Secure cable with a *strap* located within 12 inches of each steel box, eight inches from nonmetallic boxes. And let six to eight inches of cable hang out so you can easily make connections in boxes later.

4 With an electric drill and a sharp spade bit, you·can make holes in a hurry. Align them by eye, but be sure to bore at or near the center of each stud. Holes closer to the wall surface than 1¼

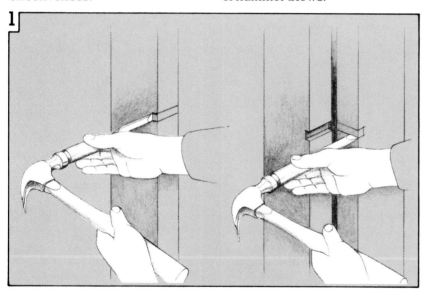

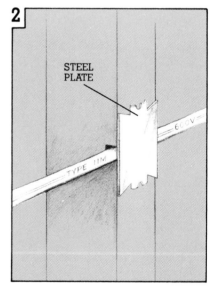

STEEL PLATE

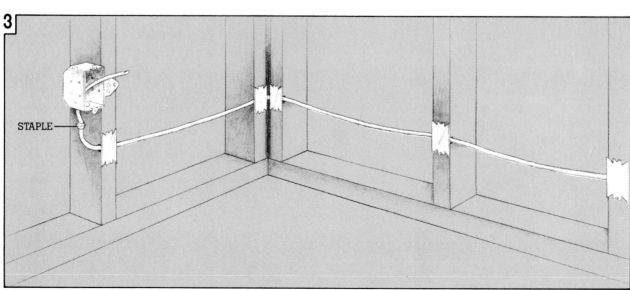

STAPLE

inches need the protection of a metal plate.

Don't holes weaken the studs? Only slightly, if at all, but to be safe use a bit just a little larger than the cable.

5 Beyond the guidelines given here, you don't need to be especially neat or exacting when running cable. Your handiwork will be covered anyway, and the current doesn't care what path it follows from one outlet to another.

Vertically Through Wall Cavities

6 Your biggest problem here is getting through the two, three, or more thicknesses of lumber at the base and/or tops of walls. A spade bit may not be long enough to do the job. If it won't penetrate, measure and bore a second hole from below or above.

Or consider investing in an electrician's bit extension, like the one shown on page 73. To use this, first bore with the spade bit alone. When you're almost to the bit's hilt, add the extension and continue. Considering the small cost of a bit extension, this is probably the best option.

A third alternative is to bore with a hand-operated brace and bit. Extensions are available for these, too.

7 On vertical runs, secure the cable with straps every 4½ feet and near changes in direction, as well as within 8 to 12 inches of boxes. You can use ordinary hammer-driven staples, too. Be careful when hammering the staples because they can damage the insulation or break a conductor. Notice here how the cable loops through the top plate to eliminate a sharp, compound bend that could conceivably cause problems. *(continued)*

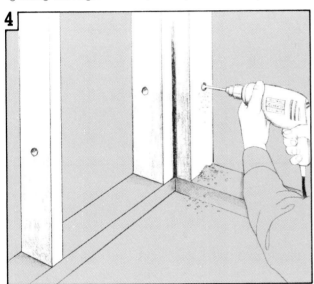

4

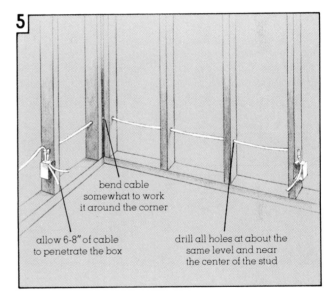

5

bend cable somewhat to work it around the corner

allow 6-8" of cable to penetrate the box

drill all holes at about the same level and near the center of the stud

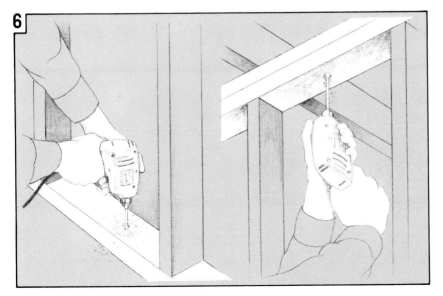

6

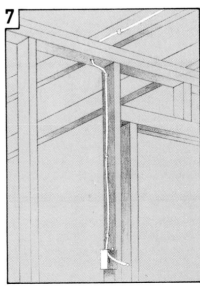

7

Running Cable in Unfinished Space *(cont.)*

Through Attics and Floors

1 What you do with cable in an attic depends on how accessible and usable the attic is. If access consists of nothing more than a simple trap door in a hallway or closet ceiling, and if headroom is too limited to consider using the space for anything, you can simply staple cable to the tops of joists. If it passes within six feet of the attic opening, however, protect the cable with 1x2 *guard strips* nailed down on either side.

2 If your attic has stairs or a permanent ladder leading to it, the National Electrical Code says you must bore holes and thread cable through the joists. Do this even if you're not planning to use the attic. The next owner of your home might decide differently.

If there's flooring up there, you'll have to pull it up to install the cable or plan an alternate route along rafters and other framing. In a basement, lighter cable should also run through holes bored through the floor joists overhead.

3 No. 8 and larger cable, which you might use for a 240-volt appliance circuit, is too stiff to thread easily through joists, so the NEC allows you to strap it to the joists. Check your local code, however, because many require that all exposed runs be protected with conduit.

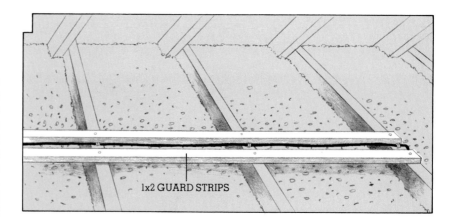

1x2 GUARD STRIPS

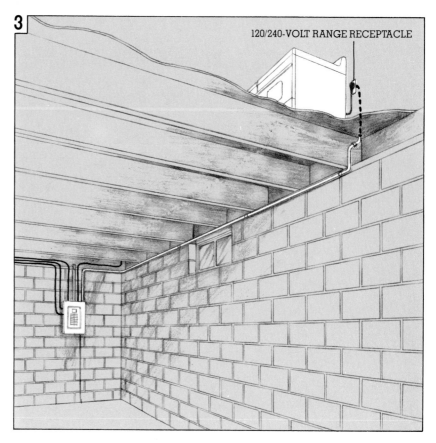

120/240-VOLT RANGE RECEPTACLE

Running Cable in Finished Space

Where There's Access From Above or Below

1 With most interior partitions you can simply drill through the top or bottom plates (you may need a bit extension), and feed cable up or down to the box opening. If you encounter fire blocking, see drawing No. 5.

2 In some cases you can reach the top or bottom of a wall from unfinished space, but can't get a drill at the plates. Then you have to work the other direction.

At the top of a wall make an opening as shown. At the bottom where there's only one plate, remove the baseboard and locate the cutout ¾ inch above the floor.

3 Start to bore through the corner of a plate, hitting it at about a 45 degree angle. Once the hole is started, tip the drill to a more vertical angle. After the bit is in deeply, pull out, install an extension, and continue on through.

4 Now push cable up or down to the box opening, and loop it through the plate. Pulling cable through walls is a two-person job. One tugs gently from the unfinished space, the other coils cable and feeds it through the box opening.

5 If there's blocking in the wall, locate it and make an opening that straddles the horizontal framing member. To get cable past, chisel a notch in the wood. Before you patch the hole, install a metal plate such as the ones shown on page 70.

If your walls are drywall and you've cut carefully, you can probably butter the edges of the cutouts with joint compound, plug them back into their openings, and tape the seams. In plaster walls, fill the opening with patching plaster. *(continued)*

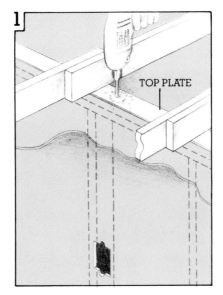

TOP PLATE

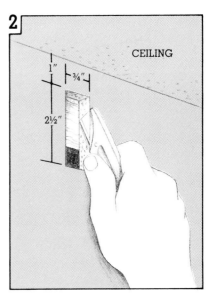

CEILING

1"   ¾"

2½"

BIT EXTENSION

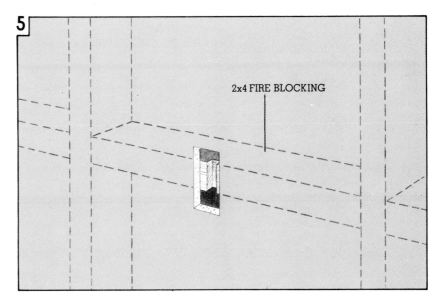

2x4 FIRE BLOCKING

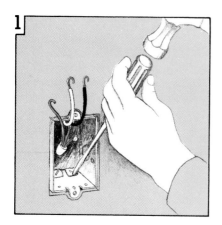

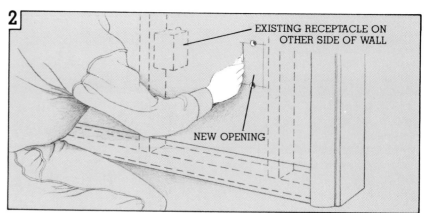

Running Cable in Finished Space *(continued)*

Fishing Cable from a Nearby Outlet

1 If you don't have access from unfinished space, don't despair. Power for a new outlet might be just around the corner or down the wall. First, go through the calculations explained on page 35 to determine whether the existing circuit can handle the new load. Next, shut off the power and check for a set of unused terminals on the receptacle you'd like to tap. If you find a pair, disconnect the device and remove a knockout, as shown here.

2 Ideally, you should situate the new outlet in the same wall cavity as the old one. If you don't, you'll have to follow a course similar to that shown on the opposite page.

3 Thread one fish tape (or a bent coat hanger) into the knockout and another through the new opening. Wiggle one or the other until you hook them together.

4 Pull the tape from the existing box though the opening and attach cable to it. If the new box doesn't have internal connectors, install an external type first, as explained on page 77.

5 Finally, pull cable to the old box and connect it. Install the new box and make your electrical hookups.

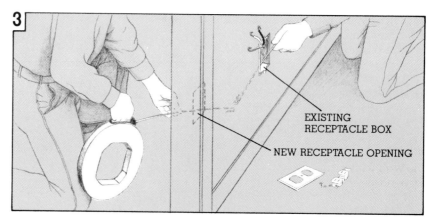

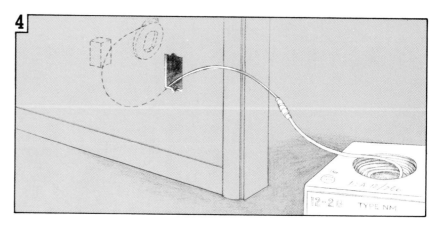

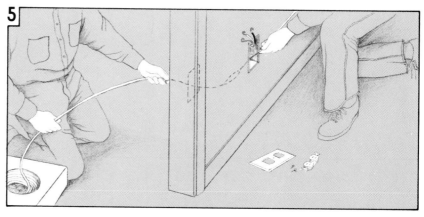

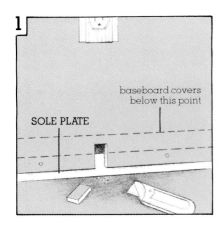

Another Way to Add Another Outlet

1 To tap an existing receptacle that's in the same room as the proposed new one, you'll have to burrow cable behind the baseboard, a job that calls mainly for patient carpentry.

The trickiest part comes first, when you have to pry trim from the wall without damaging either. Pop nails loose one at a time, wedging the wood away from the wall as you move down its length.

Once the board is free, notch wallboard or plaster beneath the outlet you'll be picking up power from. Note that when the baseboard is replaced, all scars will be covered.

2 You may find a gap at the wall's base that's big enough to accommodate the cable. If not, cut away just enough material to make a channel for it. Feed cable through a knockout in the existing box and pull it out of the wall below.

3 If your new outlet is a receptacle, you can probably just poke cable up through the wall to the opening. For a switch, you may have to fish, as shown on the opposite page.

4 To get around a door, remove its casing. Here again, there may be enough space to sneak cable through, or you may have to cut away some of the wall material. When you renail the trim, take care that you don't pierce the cable's insulation. *(continued)*

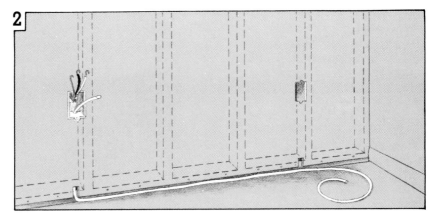

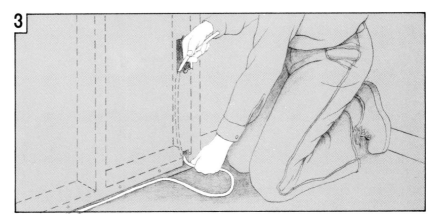

Running Cable in Finished Space *(continued)*

Running Cable for a Switch-Controlled Ceiling Fixture

1 Getting cable through a finished ceiling to a new switch outlet in a finished wall calls for planning, perseverance, and a dollop of fisherman's luck.

The luck comes first when you determine which direction the ceiling joists run. That's the path your cable will have to follow. Next, cut an opening for the new outlet in line with where the switch will go, and figure out how you're going to get power to the switch. In the situation illustrated here, a receptacle is in the same wall cavity. For other possibilities, check the preceding pages.

2 At the point where the wall and ceiling meet, make ¾x4-inch openings, then chisel a channel into the plates so the cable can pass. If the room has ceiling molding, you might be able to hide your work behind it.

3 Now slowly feed fish tape to the ceiling opening. Because it's made of springy metal, the tape tends to coil up when it meets any resistance. If that happens, pull back a few feet, shake the tape, and try again. Shaking also helps locate the tape — listen carefully and you can hear it rattling.

For longer runs you might decide to use two tapes, feeding in one from either end and snagging them as shown on page 74. This part of the job can be frustrating. Keep at it, though, and sooner or later you'll bring in a catch.

4 You can probably just push the cable down through the wall. If not, work a fish tape up, and pull the cable to the switch box opening. Staple the cable into the notch in the plates before you patch up the wall and ceiling.

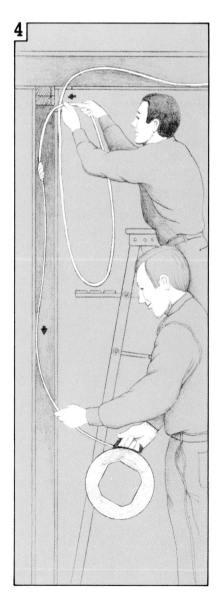

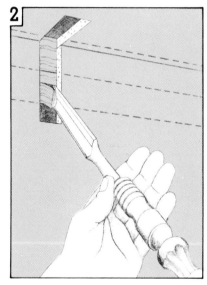

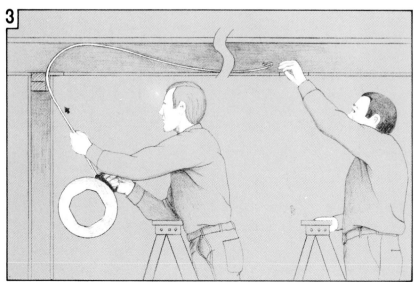

Connecting Cable to Boxes

The final step in "roughing in" an electrical installation happens when you fasten cable to the boxes. Here are the different systems you can use with Type NM wiring.

1 Slide a *clamp connector* onto the cable six to eight inches from the end and tighten its *saddle*. Then remove the *locknut*, slip the connector into a knockout, and screw the locknut back on. Finally, tighten the nut by whacking one of its lugs with a hammer and nail set.

2 A *plastic connector* is even easier to use. It snaps into the knockout, then you feed six to eight inches of cable through and secure it by turning a *capture screw*. With other types you do the same by prying up a locking wedge or squeezing the unit with a pair of pliers. A plastic connector also can be inverted and installed from inside an existing box, a big convenience in old work situations.

3 Some boxes have internal *quick clamps*. You just pry up a spring-metal tab and slip the cable through. Codes require a connector with all metal boxes.

4 Other steel boxes for non-metallic cable come with *internal saddle clamps*. Tightening a screw grips the cable.

5 Some nonmetallic boxes have a similar internal clamp. The National Electrical Code doesn't require a clamp in a nonmetallic box, provided you secure the cable within eight inches of the box. This applies only to single boxes. You need clamps with multiple units.

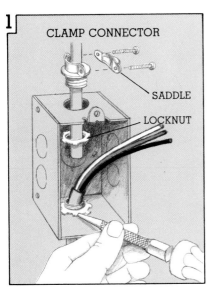

1 CLAMP CONNECTOR
SADDLE
LOCKNUT

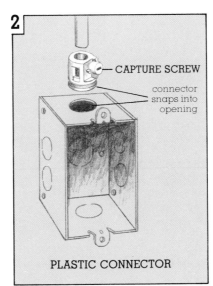

2 CAPTURE SCREW
connector snaps into opening
PLASTIC CONNECTOR

3 QUICK CLAMP

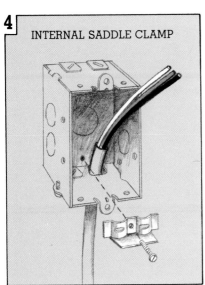

4 INTERNAL SADDLE CLAMP

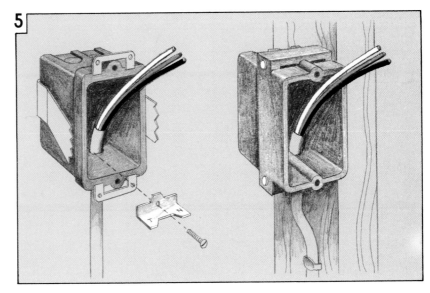

5

How to Work With Conduit

Some codes now require that conduit, not cable, be run inside the walls of a new home or addition, and almost all call for conduit in exposed locations such as unfinished garages and basements. This means that sooner or later in your electrical "career", you'll probably have to master the bending, fitting, and wire-pulling basics explained here and on pages 80-81.

Conduit has several advantages. The wires encased in the tubing can't be damaged easily. The tubing serves as a ground. And even when conduit is buried in walls, you can upgrade later by pulling new wires through.

You have to plan conduit runs carefully, however, and get the knack of bending the tubing in gentle arcs with no crimps that might impede pulling wires through.

Bending It

1 For indoor jobs, select *thin-wall electrical metallic tubing* (EMT). It's sold in 10-foot sections — ½, ¾, or 1 inch in diameter — that you shape with a conduit bender.

To get conduit around a corner, first measure from the box to the top of the bend *(distance A)*, then subtract the *take-up* that will be gained by the bend *(B)*. For ½-inch conduit allow a take-up of five inches; for ¾- and 1-inch sizes, allow six and eight inches, respectively.

2 Now slip the bender onto the tubing and align it as shown.

3 With one foot on the foot piece of the bender, pull slowly and steadily on the handle. Be careful. Tugging too sharply will crimp the tubing and you'll have to start over again with another piece.

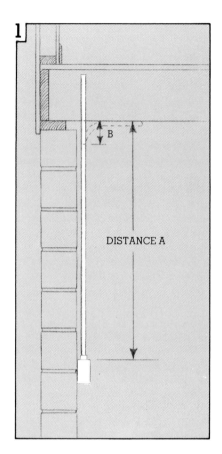

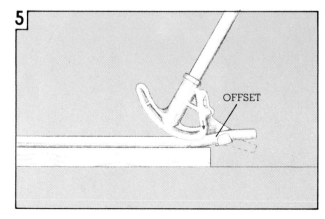

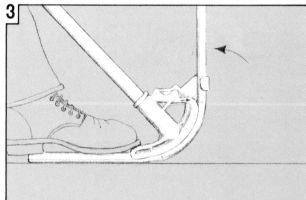

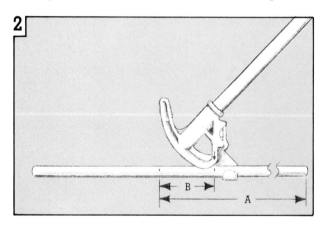

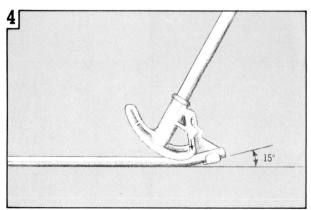

Codes don't allow crimped conduit. When the handle reaches a 45-degree angle with the floor, you've completed a 90-degree bend. Don't be surprised if your first few efforts mangle the tubing. Making smooth, crimp-free arcs takes practice.

4 If you'll be mounting conduit on a flat surface, you'll need to form *offsets* at each box. For these, first make a 15-degree bend, as shown.

5 Then flip the tubing over, move the bender a few inches farther down, and pull until the conduit is parallel with the floor. Offsets must be aligned in relation to any other bends in the tubing. With some brands, a stripe along the length of each section helps you do this.

Codes forbid a total of more than 360 degrees in bends along any run. This limits you to four 90-degree quarter-bends, three if you'll have offsets at the ends.

Changing Direction Without Bending

6 If you have more than three or four turns to negotiate, plug in a junction box. It lets you start another run. More boxes and fewer bends make wire-pulling much easier.

7 If you don't feel you want to tackle bending an offset, use an *offset connector*. These save effort, but you still must account for them in tallying up bends. Assign each a value of 45 degrees.

8 A *pulling elbow* is just the thing for negotiating corners. Its removable cover allows you to pull wires through it easily. But don't make connections inside; wires must pass through without a break.

Cutting Conduit

9 Rather than trying to compute exact take-up distances for each quarter-bend and offset, you may want to start with a section of conduit that is six or eight inches longer than the run's lineal distance, make your bends, then cut a few inches off each end.

A hacksaw makes short work of cutting conduit. To keep the tubing from rolling as you saw, hold it against a cleat, as shown. Or use a miter box, which also ensures square cuts.

A *tubing cutter* is faster yet. Clamp it onto the conduit, rotate a few times, tighten the handle, and rotate some more. A wheel-like blade slices through the metal.

10 A second, pointed blade on the cutter lets you remove sharp burrs that could chew up insulation. Stick the point into the cut tubing end and rotate the cutter. Or use a reamer or file. *(continued)*

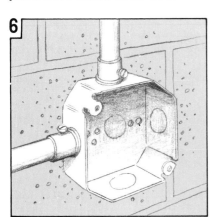

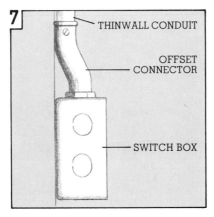

THINWALL CONDUIT
OFFSET CONNECTOR
SWITCH BOX

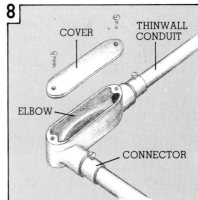

COVER
THINWALL CONDUIT
ELBOW
CONNECTOR

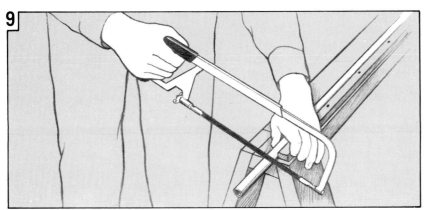

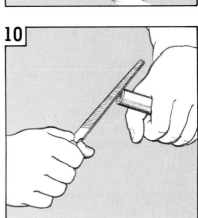

How to Work With Conduit *(continued)*

Joining and Supporting It

1 Pulling wires through conduit sometimes subjects components to stress. And because good grounding depends on secure metal-to-metal connections, you need to make sure your installation is mechanically strong. To join sections end to end, use one of the *couplings* shown here.

2 Anchor runs with a *strap* every 10 feet and within three feet of boxes. In masonry walls, use screws and plastic anchors. The barbed strap drives into framing.

To mount conduit inside walls, you can bore holes in the studs, as shown on page 70, or notch framing and protect the tubing with metal plates (see page 71). Make all connections before securing boxes to the framing.

Connecting It to Boxes

3 *Connectors* differ mainly in the way they attach to the tubing. With the *setscrew* type, you slip on the connector and tighten the screw. *Compression* connectors require a wrench; *indenters*, a special crimping tool.

All three attach to the box with the same stud and locknut arrangement used with cable connectors. You insert the stud into a knockout, turn the locknut fingertight, then hit the nut with a hammer and nail set until its lugs bite into the box.

Another version, the *90 degree angle* connector, not surprisingly makes a 90 degree connection possible. Don't forget to count it as a quarter-bend.

A *two-piece* connector comes in handy when you need to conserve space in a box. Instead of a locknut, it has a slotted compression fitting. As you tighten the nut, the fitting squeezes the conduit.

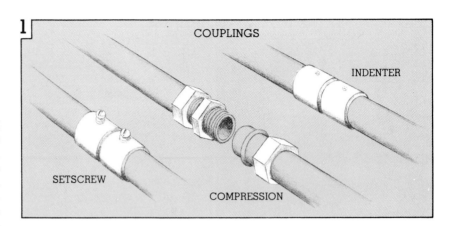

COUPLINGS
INDENTER
SETSCREW
COMPRESSION

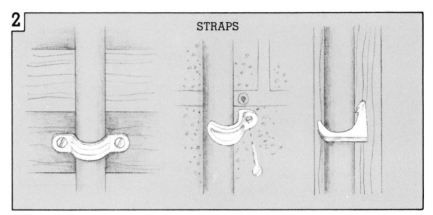

STRAPS

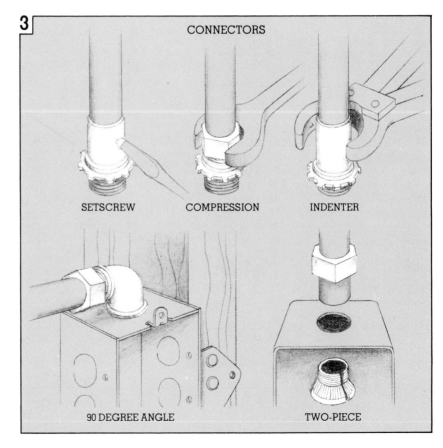

CONNECTORS

SETSCREW　　COMPRESSION　　INDENTER

90 DEGREE ANGLE　　TWO-PIECE

Pulling Wires Through Conduit

1 Now comes the moment when you realize why codes are so specific about bends, crimps, and burrs in tubing. Any hangups can lacerate insulation or even make it impossible to get wires through at all. Just how difficult your "pull" will be depends largely on how far you have to go. For a relatively short, straight run such as this, you probably can push the wires from one box to the other.

Use Type TW conductors, one black, one white, plus other colors for the hot legs of any additional circuits. Feed them carefully to protect insulation.

2 If you can't push the wires, you'll need a fish tape and a helper. Snake the tape through the conduit, then secure wires to it as shown here.

3 Now begin pulling with gentle pressure. As the wires work past bends, expect to exert more muscle. If you have lots of wires or a long pull, lubricate the wires with talcum powder or pulling grease (sold by electrical suppliers).

4 Leave six to eight inches of extra wire at each box. And never splice wires inside conduit. They must run continuously from box to box. To learn about connecting wires, see pages 86-89.

How many wires can you pull through conduit? Codes are surprisingly optimistic about this, and allow more than you can expect to stuff into a run. As a rule of thumb, you should be able to get four No. 12 wires through a short, straight section of ½-inch tubing. With bends or offsets, reduce to three or use ¾-inch conduit. Subtract one wire for each increase in wire size; add one for each increase in tubing diameter.

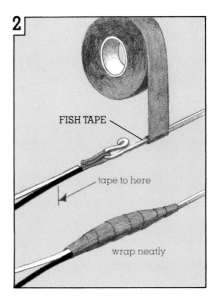

FISH TAPE

tape to here

wrap neatly

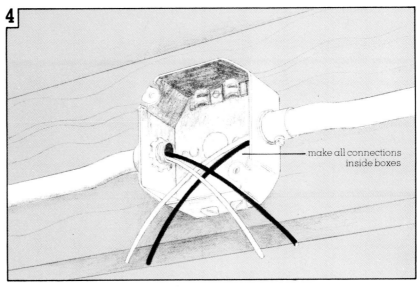

make all connections inside boxes

How to Work With Flexible Armored Cable (BX) and Flexible Metallic Conduit (Greenfield)

In many localities, armored cable and flexible conduit are orphans. You might find one or the other in an older home, but not be permitted to use it for new work. A few other communities insist that you run *only* armored, not nonmetallic cable. Check codes in your area.

Cutting Techniques

1 The first special twist you'll notice about these products is the steel covering. To get through it, hold a hacksaw at a right angle to the spirals and cut partway through the armor. With cable be especially careful not to nick insulation on the conductors inside. With Greenfield, there aren't any conductors inside.

2 Now twist the armor and it will snap free. The paper-wrapped conductors (and an aluminum *bonding strip*, if there is one) can be snipped with ordinary wire cutters. To expose the conductors for connections, cut only the sheathing at a second point about a foot away, then snap and twist off the armor.

Installation Pointers

3 and 4 Run armored cable and flexible conduit just as you would nonmetallic cable, threading it through holes or plate-protected notches in studs and other framing members. BX and Greenfield are much heavier and stiffer than NM, so you need to change directions gradually.

Like NM, armored cable usually is used only in concealed loca-

tions, but you can take advantage of its flexibility in exposed runs of up to 24 inches. Furnace motors often are connected this way to compensate for vibration.

5 Support metal-clad cable with *straps* or *staples* every 4½ feet and within 12 inches of boxes. If you're fishing through existing walls or ceilings, you won't be able to do this, of course, but secure the run as best you can.

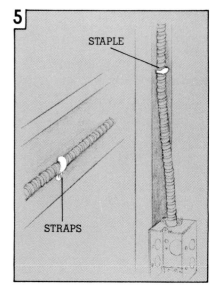

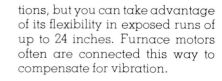

BONDING STRIP

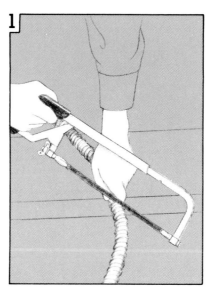

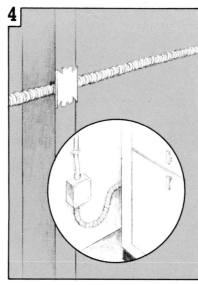

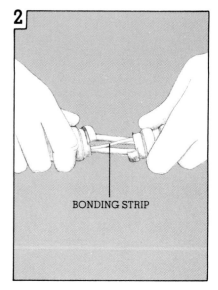

STAPLE

STRAPS

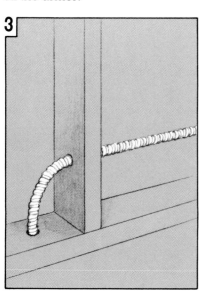

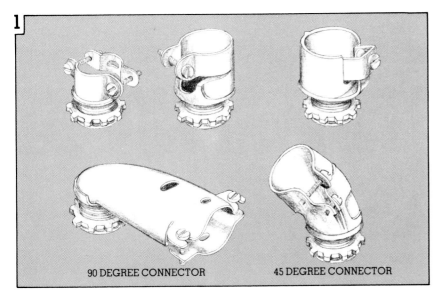

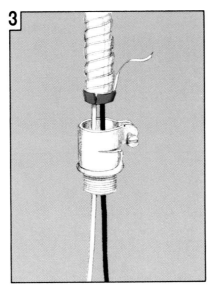

90 DEGREE CONNECTOR   45 DEGREE CONNECTOR

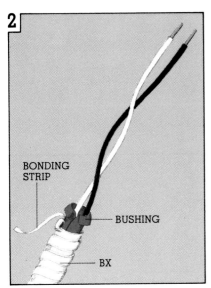

BONDING
STRIP

BUSHING

BX

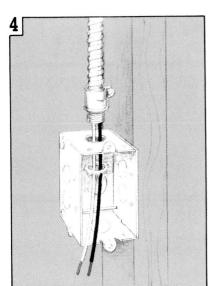

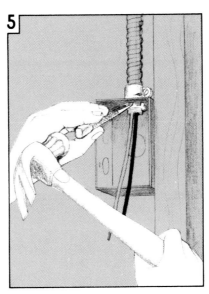

Connecting BX and Greenfield to Boxes

1 The important thing to remember when you're hooking up to armored cable or flexible metal conduit is that the wires must be protected against the armor's sharp edges.

One way to safeguard the wires is to install an *anti-short bushing*, as shown in sketch 2. With these you use *peep-hole* connectors that let inspectors verify that there's a bushing inside. If you'd rather not bother with bushings, spend a bit more and buy *insulated-throat* connectors (not shown). *Right-angle* and *45-degree* connectors come insulated and non-insulated.

2 When you pull off the brown paper that surrounds the wires in BX cable, try to rip it back an inch or so inside the armor. This leaves room to slip in a bushing. If the cable you're using has a bonding strip, fold it back as shown.

3 Now just slip on a connector and secure the bonding strip to the connector's tightening screw. Check to be sure the bushing is in place, then tighten the screw. As with rigid conduit, flexible conduit and armored cable are self-grounding. You don't need a third wire with them.

4 Slip the wires and connector into a knockout, slip on a locknut, and turn it fingertight. As with all wiring, connections can be made only in boxes.

5 Draw the locknut snug with a hammer and nail set. Finally, tug on the cable to be sure everything is securely fastened.

Your Switch and Receptacle Options

If you've always thought that a switch is a switch and a receptacle is a receptacle, prepare for a surprise. The array illustrated here represents only a sampling of the dozens of different UL-approved devices offered by electrical equipment manufacturers.

Some of the alternatives are strictly color choices. Besides brown and ivory, most also come in white, and some in black, gray, even red, yellow, blue, and other hues, with wall plates to match.

Many of the differences are more than decorative, however. Let's take a look at them.

1 For most of your switching needs, you'll probably choose a *single-pole toggle*. Standard toggles, the ones that flip on and off with a loud snap, are all but obsolete in home wiring installations. Today's *quiet* switches operate with a scarcely audible click. And *silent* switches, some of them illuminated so you can find them in the dark, turn on and off with no noise whatsoever.

Three- and *four*-way toggle switches are available in quiet and silent versions. To learn about wiring these, see pages 41-44. *Push* switches come with one or two buttons. The single-button version is a quiet type; the two-button is silent.

Looking for something a bit more decorative? Select a *rocker*. Keep in mind, however, that you'll pay a premium for their sleekness. *Tamperproof* switches can only be operated with a key. These make special sense for shop tools and other devices you don't want kids to play with.

Dimmers let you adjust lighting levels to suit your mood and needs. The *rotary* type comes in two versions. One turns lights on or off with a push. The other does the same when you turn the control fully counterclockwise. Just a tap of your fingers operates a *touch* dimmer; holding them on the switch a little longer adjusts the brightness. More about dimmers on pages 32 and 33.

To add another switch in a single box, get a *double*. It takes up no more space than a standard receptacle.

2 Have you ever noticed that the left and right slots in a receptacle are different sizes? That's because plugs on some appliances are *polarized* so they'll be grounded through the neutral side of the circuit. Standards by Underwriters' Laboratories and the National Electrical Code now require that lamp cords also be polarized.

Duplex receptacles let you plug two items into the same outlet. Many newer homes have *20-amp grounded* types. With aluminum wiring, use only *15-amp* types labeled *CO/ALR*. If your outlet boxes aren't grounded, install *15-amp ungrounded* receptacles. These are made only for replacements in existing circuits.

The switch in a combination *switch/receptacle* can be hooked up to control the receptacle it's paired with or another outlet elsewhere. A *20-amp single* makes it all but impossible to overload a critical circuit. Get a *twist-lock* version and an appliance can't be accidentally disconnected, either.

Select *240-volt* receptacles according to the appliance's amperage rating. Plugs for appliances of 15, 20, 30, and 50 amps have different blade configurations.

1 SINGLE-POLE TOGGLE
THREE-WAY
FOUR-WAY

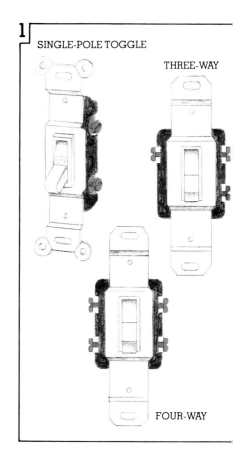

2 20-A GROUNDED
15-A GROUNDED

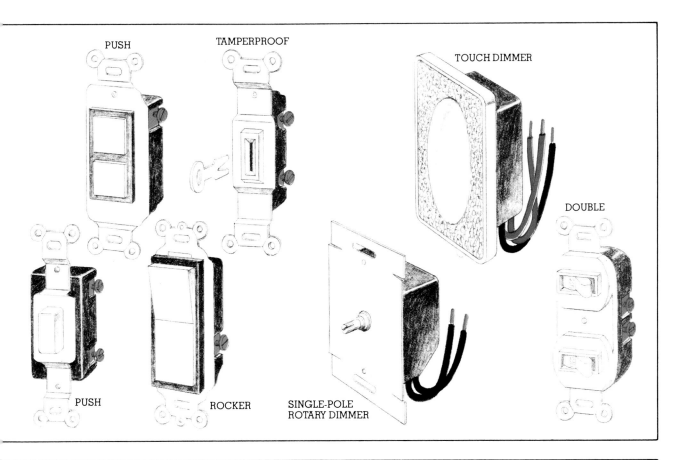

PUSH

TAMPERPROOF

TOUCH DIMMER

DOUBLE

PUSH

ROCKER

SINGLE-POLE
ROTARY DIMMER

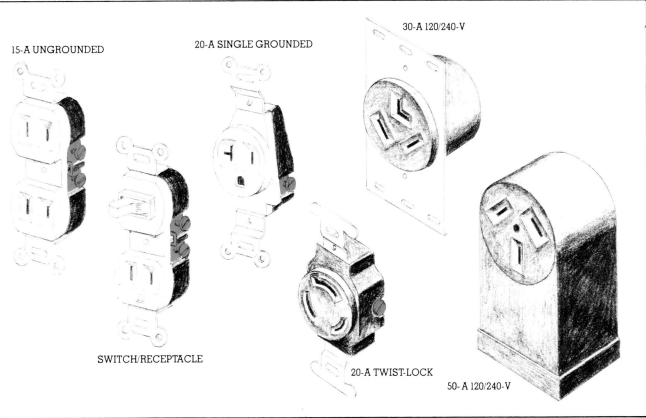

15-A UNGROUNDED

20-A SINGLE GROUNDED

30-A 120/240-V

SWITCH/RECEPTACLE

20-A TWIST-LOCK

50- A 120/240-V

Making Electrical Connections

The final (and most fun) phase of an electrical installation comes when you tie all the wires you've run to each other and to the switches, light fixtures, and receptacles they'll be supplying. Here's how to strip away insulation and make the right connections.

Stripping Cable

1 The easiest way to remove the plastic sheath from nonmetallic sheathed cable is with an inexpensive *cable ripper*. (You may find it easier to strip the cable before connecting it to the box.) Slip six to eight inches of cable into the ripper's jaws, squeeze, and pull. This slits open the sheathing without damaging insulation on the conductors inside.

2 Now peel back the sheathing and any paper or other filler material. You'll find two or three separately insulated conductors and a bare ground.

3 Cut off the insulation and paper with a utility knife. Nicking the insulation on the conductors could cause a short, so always cut away from them. Leave at least ¼ inch of sheathing inside the outlet box.

4 You can also strip sheathing with a utility knife. Make a shallow, lengthwise cut down the center, as shown. Again, take care to avoid cutting into the conductor insulation. Remove the sheathing and paper as shown in sketch 3.

To learn about removing the sheathing from armored cable, see page 82. Single-conductor wires that run inside conduit have no sheathing. With these you strip the conductors as shown in sketches 5 and 6.

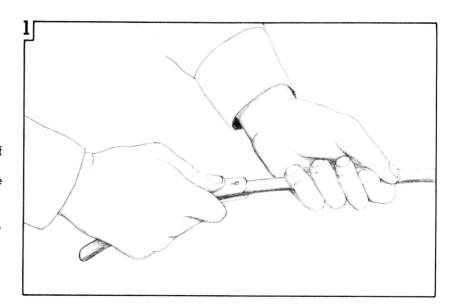

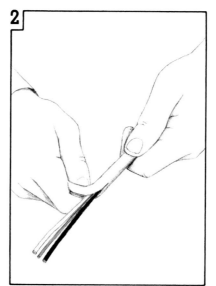

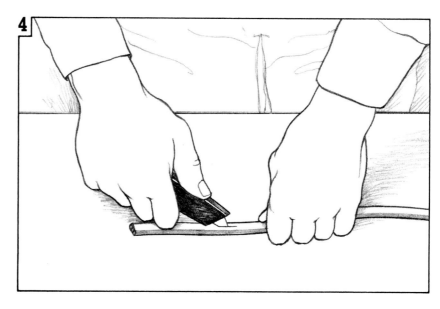

5 A *combination tool* makes short work of insulation removal. Locate the size of the wire on its jaws, clamp down, pull the tool away from you, and the covering pops off. Some cable rippers can strip in much the same way.

6 You can also remove insulation with a sharp knife. Start the cut at an angle so you don't nick the wire, then peel off the insulation. You'll have to use this technique to strip wires larger than No. 6.

Connecting Wires to Each Other

7 To splice two or more wires, a *solderless connector* is your best bet. Some screw on, others must be crimped (use the crimping jaws of your combination tool), still others have a setscrew and threaded sleeve. Check your code to learn which type is preferred in your community.

With all but the setscrew variety, hold the wires side by side, twist them together, and turn or crimp on a connector. Make sure that no bare conductors are exposed and that all wires are locked in.

Solderless connectors come in a variety of color-coded sizes to suit the gauges and numbers of wires you'll be splicing.

8 A few codes require that all splices be soldered. Others prohibit soldering house wiring. Again, you start by twisting together the wires.

9 Next, heat them with a soldering iron and melt rosin-core solder over the splice. Soldering requires some practice, but once you master the knack of flowing on just enough to do the job, it goes quickly.

10 After the solder cools, wrap the splice with electrical tape. Cover about an inch of insulation as well as the bared conductors. Some electricians also tape solderless connectors. *(continued)*

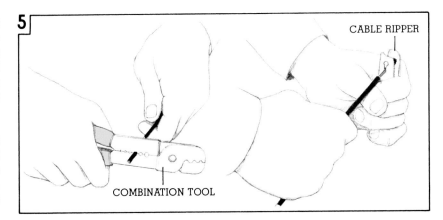

5 CABLE RIPPER / COMBINATION TOOL

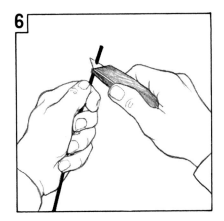

6

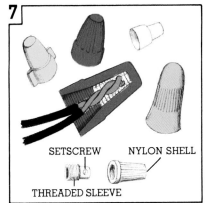

7 SETSCREW   NYLON SHELL / THREADED SLEEVE

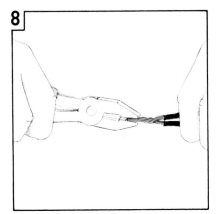

8

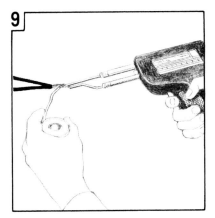

9

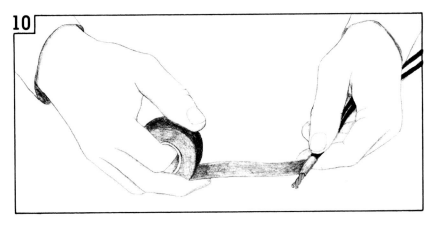

10

Connecting Wires to Each Other *(continued)*

11 If you attempt to splice a stranded wire to a solid conductor, as you might have to when you're hooking up a light fixture or dimmer switch, you'll discover that the more flexible stranded material bunches up when you try to screw on a connector. To make a more secure connection, first wrap the stranded wire around the solid one, as shown.

12 Then bend the solid conductor over the wrap. Finally, turn on a solderless connector and tape any wire that shows.

Connecting Wires to Switches and Receptacles

1 Many of today's receptacles and switches offer a choice of connections. You can strip wires and push them into holes in the rear of the device, or wrap the conductors around screw terminals on the sides.

Push-ins have a slight edge in convenience. On the back of the device you'll find a *strip gauge* that indicates exactly how much insulation to remove.

2 Insert the bare wires into the holes and push until a metal spring grips them. With receptacles, holes on the neutral side will be marked *white*. Because switches connect only to hot wires, it doesn't matter which holes you use.

3 To disconnect a push-in, insert stiff wire or the blade of a small screwdriver into the slot next to the terminal. If you've inadvertently bared more wire than necessary, either release the wire and snip off the extra, or, if you'd rather, wrap the exposed section with electrical tape.

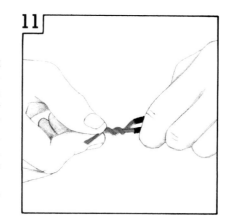

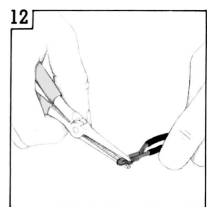

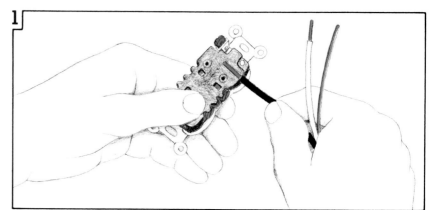

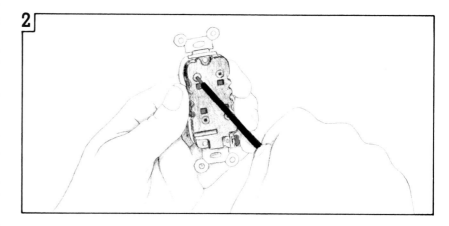

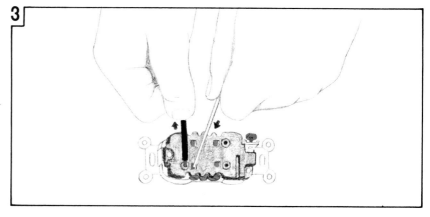

4

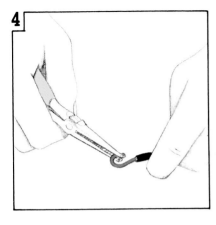

5

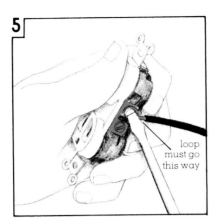

loop
must go
this way

6

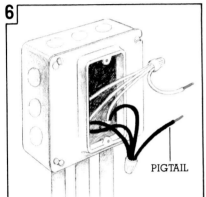

PIGTAIL

7

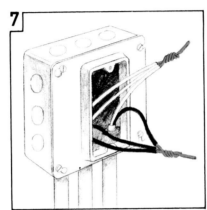

8

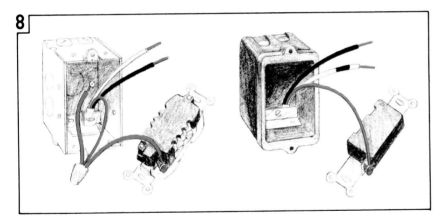

9

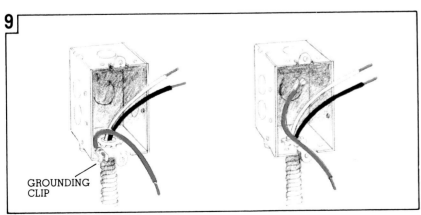

GROUNDING
CLIP

4 With screw terminal-type devices, you bare just enough wire to wrap around the terminal, then form it into a loop with long-nose pliers. It takes practice to make loops that are neither too big nor too small and that lie flat.

5 Always hook the wire clockwise around the terminal so that tightening the screw will close the loop. With receptacles, the black wires go to the brass side, white to silver. Don't overtighten the screws, because if you crack a device, you have to throw it out.

6 Never attach more than one wire to a terminal. Codes don't allow it. Instead, you can connect two or more wires with a third short one to make the *pigtail* shown here. In computing the permitted number of wires in a box (see page 62), count each pigtail as one.

7 Or you can twist the wires for this special splice. It must be soldered and taped, and also gets a value of one in your wires-in-the-box calculations.

8 How you ground receptacles and switches depends on the type of wiring you're using as well as the type of box. If you're working with nonmetallic sheathed cable and metal boxes, use the arrangement shown in the left-hand portion of the sketch. With nonmetallic boxes, the cable's grounding wire connects directly to the device, as shown at right.

9 With armored cable and flexible or rigid conduit, the metal covering serves as the grounding wire. For these, just run a green or bare wire from the device's grounding screw to a screw in the box, or secure it to the box with a *grounding clip*. More about grounding on pages 9, 21, 46, 47, and 59.

PRODUCTS YOU MAY WANT TO TRY

Looking for ways your family can use electricity safely, conveniently, and efficiently? Here's a sampling of handy accessories, most of which are available at local hardware stores, home centers, or electrical supply houses.

Screw in a *light-activated switch* and any ordinary lamp will turn on at dusk, off again at dawn. This version has a swiveling sleeve that enables you to point the unit's photocell toward a window or other source of natural light.

Timers offer another way to control lamps and appliances automatically. You can program these 24-hour clocks to turn devices on and off one or sometimes several times each day. For a battery charger, air conditioner, or other high-amperage appliances, get a heavy-duty timer.

The *dimmers* shown on pages 32 and 33 must be permanently installed. Not so with this *tabletop model* (G.E. "Table Top Dimmer"). It's as portable as any lamp. Just plugging into a receptacle puts full-range lamp intensity control at your fingertips.

Tired of replacing fuses? A *miniature circuit breaker* saves the bother. You just screw it into an ordinary fuse socket. To reset it, push the red button.

Raceway makes it easy to get electricity from one point to another without ripping into finished walls and ceilings. Modular components let you tap into an existing outlet, then run wires through plastic or steel channels to surface-mounted receptacles, switches, or light fixtures.

If it seems you're always in need of another outlet, invest in a *receptacle adapter*. This handy item can double or even triple the capacity of a receptacle. Take care, though, that you don't overload the circuit. The type shown here works only for low-amperage, ungrounded applications. Others accept grounded plugs.

Receptacles and plugs down near a toddler's eye level constitute a definite safety hazard. *Child-safe inserts* and *safety receptacles* keep little fingers out of trouble. With the receptacles, you twist plugs in and out. Install them as shown on pages 20 and 21.

This *extension cord reel* ("Bench Boy™", Fleck Mfg. Co., Tillsonburg, Ontario), offers several advantages over ordinary cords. It has a duplex receptacle so you don't have to unplug one tool every time you need to use another. The unit also includes a 10-amp circuit breaker that saves trips to the service panel if you overload it or if a tool malfunctions. And when you're done working, just crank in the cord. *(continued)*

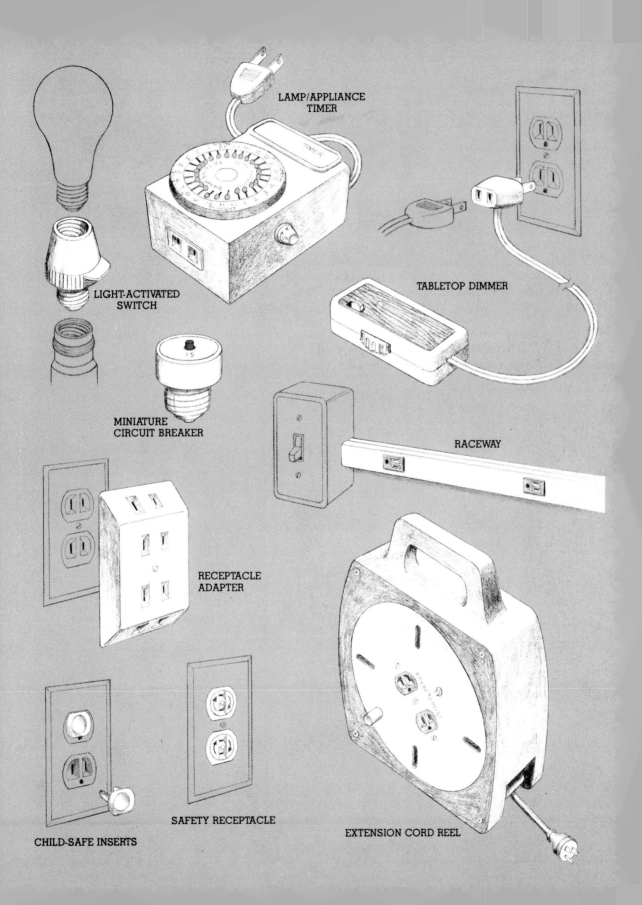

LAMP/APPLIANCE TIMER

LIGHT-ACTIVATED SWITCH

MINIATURE CIRCUIT BREAKER

TABLETOP DIMMER

RACEWAY

RECEPTACLE ADAPTER

CHILD-SAFE INSERTS

SAFETY RECEPTACLE

EXTENSION CORD REEL

Products You May Want to Try *(continued)*

Have you ever settled into bed only to realize you left the kitchen lights blazing? Or, wondered what to do about a suspicious noise at the back door? Or wished you could start the coffee perking before you arrive downstairs in the morning? If so, you'll appreciate the electronic virtues of this remote-control switching system (BSR "X-10 Control System").

It lets you operate up to 16 lights and appliances from a single station. You just plug a *command module* into any wall receptacle, then add *lamp, switch,* or *appliance modules* for each device you'd like to control.

Pushing the command module's buttons sends signals through your home wiring to turn lights and appliances on or off, dim or brighten lights, even switch everything in the system on or off.

Have you ever wondered exactly how much electricity your household uses at a given moment? An *energy monitor* ("Fitch Energy Monitor", Fitch Creations, Inc., Carrboro, NC 27510) lets you know by a digital readout that reports in cents per hour what you're spending. Just program in your utility's electric rate and the monitor does the rest.

It works by measuring the flow of current through your home service panel, then using calculator circuitry to convert to the cents-per-hour figure.

The monitor installs in a standard, double-gang wall box. You need to run a 120-volt line to the unit's transformer, and low-voltage wires to the service panel. There, a pair of current detectors straddle your home's main power cables. The manufacturer suggests hiring an electrician to connect these.

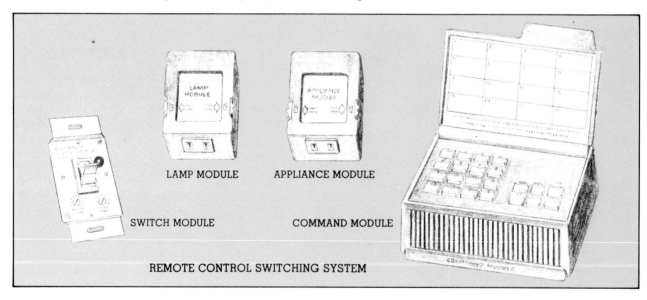

LAMP MODULE   APPLIANCE MODULE

SWITCH MODULE   COMMAND MODULE

REMOTE CONTROL SWITCHING SYSTEM

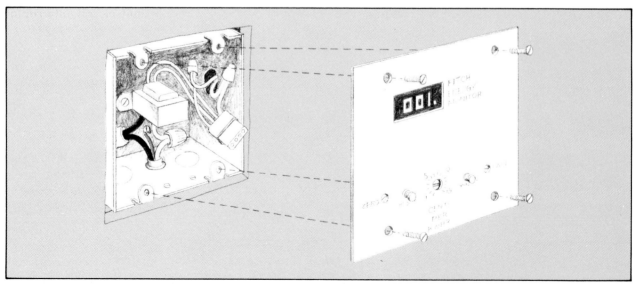

Glossary

Occasionally puzzled by electrical terminology? If so, these definitions should help. For words not included here, or for more about those that are, refer to the index, which begins on page 95.

Amp (A). A measurement of the amount of electrical current in a circuit at any moment. See *volt* and *watt*.

Armored cable. Two or more insulated wires wrapped in a protective metal sheathing.

Ballast. A *transformer* that steps up the voltage in a fluorescent lamp.

Bimetal. Two metals that heat and cool at different rates to open or close a circuit automatically. They are commonly used in circuit breakers and thermostats.

Box. A metal or plastic enclosure within which electrical connections are made.

Bus bar. A main power terminal to which circuits are attached in a fuse or breaker box. One bus bar serves the circuit's *hot* side, the other the *neutral*.

BX. A trade name for *flexible armored cable*.

Cable. Two or more insulated *conductors* wrapped in metal or plastic sheathing.

Circuit. The path of electrical flow from a power source through an *outlet* and back to *ground*.

Circuit breaker. A switch that automatically stops electrical flow in a circuit in case of an overload or short.

Codes. Local laws governing safe wiring practices. See *National Electrical Code*.

Conductor. A wire or anything else capable of carrying electricity's energy.

Conduit. Rigid or flexible tubing through which you can run wires.

Contact. The point where two electrical conductors touch.

Continuity tester. A device that tells whether a circuit is capable of carrying electricity.

Delayed-start tube. A type of fluorescent tube that takes a few seconds to warm up.

Dimmer. A switch that lets you vary a light's intensity.

Duplex receptacle. A device that includes two plug outlets. Most of the receptacles in your home probably are duplexes.

Electrical metallic tubing (EMT). Thinwall rigid conduit suitable for indoor use.

Electrons. Invisible particles of charged matter, moving at the speed of light through an electrical circuit.

Fishing. The process of getting cables through finished walls and ceilings.

Fish tape. A long strip of spring steel used for fishing cables and for pulling wires through conduit.

Fixture. Any light or other electrical device that is permanently attached to a home's wiring.

Flexible metal conduit. Tubing that can be bent easily by hand. Also known as *Greenfield*.

Fluorescent tube. A light source that uses an ionization process to produce ultraviolet radiation. This turns into visible light when it hits a coating on the tube's inner surface.

Four-way switch. A type of switch used to control a light from three or more locations.

Fuse. A safety device designed to stop electrical flow if a circuit shorts or is overloaded. Like a *circuit breaker*, a fuse protects against fire from overheated wiring.

Ganging. Assembling two or more electrical components into a single unit. Boxes, switches, and receptacles often are ganged.

General-purpose circuit. Serves a number of light and/or receptacle outlets. Also see *small-appliance* and *heavy-duty circuits*.

Ground. Refers to the fact that electricity always seeks the shortest possible path to the earth. Neutral wires carry it to ground in all circuits. An additional grounding wire, or the sheathing of metal-clad cable or conduit, protects against shock from a malfunctioning tool or other device.

Ground fault circuit interruptor (GFCI). A safety device that senses any shock hazard and shuts down a circuit.

Heavy-duty circuit. Serves just one 120-240-volt appliance. Also see *general-purpose* and *small-appliance circuits*.

Hot wire. The conductor that carries current *to* a receptacle or other outlet. Also see *neutral wire* and *ground*.

Incandescent bulb. A light source with an electrically charged metal filament that burns at white heat. *(continued)*

Glossary (continued)

Insulation. A non-conductive covering that protects wires and other electricity carriers.

Junction box. An enclosure used for splitting circuits into different branches. In a junction box, wires connect only to each other, never to a switch, receptacle, or fixture.

Kilowatt (kw). One thousand watts. A *kilowatt hour* is the standard measure of electrical consumption.

Knockouts. Tabs that can be removed to make openings in a box. The openings accommodate cable and conduit connectors.

Leads. Short wires.

National Electrical Code (NEC). A set of rules governing safe wiring methods drafted by the National Fire Protection Association. Local codes sometimes differ from and take precedence over NEC requirements.

Neon tester. A device with two leads and a small bulb that determines whether a circuit is carrying current.

Neutral wire. The conductor that carries current *from* an outlet back to ground (it's always clad in white). Also see *hot wire* and *ground*.

Nonmetallic sheathed cable. Two or more insulated conductors clad in a plastic covering.

Outlet. Any potential point of use in a circuit. Receptacles, switches, and light fixtures are all considered outlets.

Overload. A condition that exists when a circuit is carrying more amperage than it was designed to handle. Overloading causes wires to heat up, which in turn blows fuses or trips circuit breakers.

Polarized plugs. Plugs designed so the hot and neutral sides of a circuit can't be accidentally reversed. One blade of the plug is wider than the other.

Rapid-start tubes. Fluorescent tubes that light up almost instantly.

Receptacle. An outlet that supplies power for lamps and other plug-in devices.

Rigid conduit. Wire-carrying tubing that can be bent only with a special tool.

Romex. A trade name for *nonmetallic sheathed cable*.

Service entrance. The point where power enters a home.

Service panel. Your home's main fuse or breaker box.

Short circuit. A condition that occurs when hot and neutral wires contact each other. Fuses and breakers protect against fire, which can result from a short.

Small-appliance circuit. Usually has only two or three 20-amp receptacle outlets.

Solderless connectors. Screw-on or crimp-type devices that join two wires.

Stripping. Removing insulation from a wire.

Sub-panel. A smaller, subsidiary fuse or breaker box.

System ground. A wire connecting a service panel to the earth. It may be attached to a main water pipe or to a rod driven into the ground.

Three-way switch. Operates a light from two locations.

Time-delay fuse. A fuse that does not break the circuit during the momentary overload that can happen when an electric motor starts up. If the overload continues, this fuse blows as does any other.

Transformer. A device that reduces or increases voltage. In home wiring, transformers step down current for use with low-voltage equipment such as thermostats and door chime systems.

Travelers. Two of the three conductors that run back and forth between switches in a three-way installation.

Underwriters' Laboratories (UL). An independent testing agency that examines electrical components for possible safety hazards.

Volt (V). A measure of electrical potential. Volts x *amps* = *watts*.

Voltmeter. A device that measures voltage in a circuit and performs other tests.

Wall box. A rectangular enclosure for receptacles and switches. Also see *junction box*.

Watt (W). A measure of the power an electrical device consumes. Also see *volt*, *amp*, and *kilowatt*.

Index

A-B

Amperage, 8
  ratings, 56-57
  and receptacles, 84
Attics:
  fans, installing, 52-53
  running cable in, 72
BX (flexible armored cable), 61
  working with, 82-83
Bathroom fans, installing, 51
Boxes, electrical, 62-67
  connecting conduit and cable
    to, 77, 80, 83
  "ganging" of, 40, 62
  installing:
    in finished space, 66-67
    in unfinished space, 64-65
  kinds, 62
    and grounding
      arrangements, 89
  size needed, determining, 63
  table, 62
  See also Junction boxes
Breaker boxes, 12
  anatomy of, 13
  See also Service panels
Breakers: see Circuit breakers

C

Cable ripper, 10, 86
Cable:
  connecting to boxes, 77, 83
  flexible armored (BX), working
    with, 82-83
  paths for, 68
  running:
    armored (BX), 82
    in finished space, 68, 73-76
    in unfinished space, 70-72
  stripping, 86
  types of, 61
Chimes, door, repairing, 26-27
Circuit breakers, 8, 12
  ground fault circuit interrupter
    (GFCI), 47
  miniature, 90
  single-pole versus two-pole,
    59
  troubleshooting, 13

Circuits, 6
  adding, 56-59
    need for permit, possible, 5
  categories of, 56
  de-rating calculations, 57
  grounding, 9, 21, 84, 89
  load on, determining, 35
  low-voltage, 26, 54
  monitoring, with GFCIs, 46-47
  need selector, 56
  overloads and shorts, 8
    correcting, 13
    and fuses, 14, 15
  tapping into, locations for, 35
Codes, electrical, 5, 60
  conduit regulations, 78, 81, 82
  See also National Electrical
    Code
Combination tool, 10, 87
Conduit, 78
  bending, 78
    alternatives, 79
  connecting to boxes, 80, 83
  cutting, 79, 82
  flexible metallic (Greenfield),
    61
    working with, 82-83
  installing, 80, 82
  outdoor use, 48, 50
  pulling wires through, 81
  tools for, 10, 78, 79
  types, 61
Connectors, types of, 77, 80,
  83, 87
Cords:
  extension cord reel, 90
  kinds of, 16
  replacing, 19

D-G

Dimmer switches, 84
  installing, 32
    fluorescent, 33
    three-way, 32, 42
  tabletop model, 90
Door chimes, repairing, 26-27
Door latch, electrically
  controlled, 55
Electrical metallic tubing,
  thinwall (EMT), 61, 78
Electrons, flow of, 6
  rate, 8
Energy monitor, 92
Extension cord reel, 90
Fans, installing:
  attic, 52-53
  bathroom, 51

Fire, possibility of, 5, 8
Fish tape, use of, 74, 76, 81
Fluorescent fixtures, 24-25
  dimmer switches for, 33
Fuse puller, 10, 15
Fuses and fuse boxes, 8, 14-15
  miniature circuit breakers,
    substitution of, 90
  See also Service panels
Greenfield (flexible metallic
  conduit), 61
  working with, 82-83
Ground fault circuit inter-
  rupter, 9, 46
  installing, 47
Grounding, 9
  and receptacles, 9, 21, 84, 89

I-N

Insulation:
  removing, 87
  repairing, 13
Insurance policies, and owner-
  installed wiring, 5
Intercoms, installing, 54-55
Junction boxes, 38, 62, 68, 79
  installing, 65
  locating, 35
LB fittings, 48, 50
Lamps and light fixtures:
  boxes, 62
    installing, 65, 67
  components:
    fluorescent, 24
    incandescent, 22
    mounting systems, 31
    standing lamps, 28, 29
    swag lamps, 28, 30
  high-wattage bulbs in, 13
  installation:
    boxes, 65, 67
    with multiple switches, 40-
      44
    post lamps, outdoor, 49
    beyond switch, 39, 42
    with switches beyond, 38,
      43
  mounting, 31
  sockets, 16
    replacing, 19
  switches: see Switches
  troubleshooting:
    fluorescent, 24-25
    incandescent, 22
  wiring:
    standing lamps, 29
    swag lamps, 30

National Electrical Code (NEC)
   requirements, 5, 60
   about boxes, 77
     sizes, table of, 62
   cable, running, in attics, 72
   circuit needs, table of, 56
   lamp cords, polarization of, 84
   receptacles, spacing of, 64
National Fire Protection
   Association, 5

O-R

Offsets, in conduit, forming, 78
   alternative, 79
Outdoor wiring, 48-50
Outlets:
   definition of, 6
   See also Boxes; Receptacles;
     Switches
Overloads, 8
   correcting, 13
   and fuses, 14, 15
Photocells, controlling lights
   with, 49, 90
Pigtails, making, 89
Pliers, 10
Plugs:
   kinds of, 16
   replacing, 17-18
     need for, 16
Polarization, 84
Post lamps, installing, 49
Raceway, 90
Receptacle analyzer, 10
Receptacles, 6
   adapters, 90
   adding, by tapping into, 36
     outdoors, 48
     running cable, 74-75
   boxes, 62
     installing, 64, 66
   child-safe, devices for, 90
   connecting wires to, 88-89
   exterior, 48, 50
   grounding of, 9, 21, 89
   kinds of, 84
   replacing, 21
     with GFCI type, 47
   splitting, 37
Remote-control switching
   system, 92

S

Sensors:
   photocells, 49, 90
   smoke detectors, 45
Service panels, 6, 8
   adding circuits to, 57, 59
   breaker boxes, 12-13
   and cutting of power, 8, 20, 58
   fuse boxes, 14-15
   and grounding, 9
   replacement products, 47, 90
Shocks, electrical, 9
Short circuits, 8
   correcting, 13
   fuses blown by, 14
Smoke detectors, wiring, 45
Soldering, 10, 19, 87
Switches:
   boxes, 62
     installing, 64, 66
   connecting wires to, 88-89
   dimmer: *see* Dimmer switches
   grounding, 89
   installing:
     boxes, 64, 66
     for control of receptacles, 37
     dimmer, 32-33, 42
     beyond fixtures, 38, 43
     with fixtures beyond
       switches, 39, 42
     four-way, 44
     photocells, 49
     running cable, 76
     separate, for two fixtures,
       40
     three-way, 32, 41-43
   kinds of, 84
   light-activated, 49, 90
   remote-control system, 92
   replacing, 20
   testing, 23

T-Z

Terminals, switch and
   receptacle:
   push-in, 88
   screw, 89
Testers, 10
   use of, 20, 21, 23
Thermostats:
   attic fans controlled by, 53
   timed-setback, installing, 34
Timers, 90
Tools for electrical work, 10
   combination tool, 10, 87
   for conduit and cable, 10, 78,
     79, 86

Tools for electrical work
   (continued)
   drills and bits, 10, 71, 73
   fish tape, use of, 74, 76, 81
   testers, use of, 10, 20, 21, 23
   voltmeter, 10, 27
Track lighting, 31
Transformers:
   for door chimes, 26
     testing, 27
   for intercoms, 55
Tubing cutter, 79
Underwriters' Laboratories
   symbol, importance of, 5
Ventilating fans, installing,
   51-53
Voltage, 8
Voltmeter, use of, 10, 27
Wattage, 8
   in de-rating calculations, 57
Wire, 6
   aluminum, 61
   connecting to terminals, 88-89
   grounding wires, 9
   pulling through conduit, 81
   removing insulation, 87
   repairing insulation, 13
   running underground, 48-50
   sizes, 19, 61
   splicing, 87-88, 89
   See also Cable

If you would like to order any additional copies of our books, call 1-800-678-2802 or check with your local bookstore.
